Flood Hazard Identification and Mitigation in Semi- and Arid Environments

Flood Hazard Identification and Mitigation in Semi- and Arid Environments

editors

Richard H French
University of Texas at San Antonio, USA

Julianne J Miller
Desert Research Institute, USA

World Scientific

NEW JERSEY · LONDON · SINGAPORE · BEIJING · SHANGHAI · HONG KONG · TAIPEI · CHENNAI

Published by

World Scientific Publishing Co. Pte. Ltd.

5 Toh Tuck Link, Singapore 596224

USA office: 27 Warren Street, Suite 401-402, Hackensack, NJ 07601

UK office: 57 Shelton Street, Covent Garden, London WC2H 9HE

British Library Cataloguing-in-Publication Data
A catalogue record for this book is available from the British Library.

**FLOOD HAZARD IDENTIFICATION AND MITIGATION IN SEMI-
AND ARID ENVIRONMENTS**

ISBN-13 978-981-4355-09-4
ISBN-10 981-4355-09-7

Typeset by Stallion Press
Email: enquiries@stallionpress.com

Printed in Singapore.

Foreword

Alluvial fans are ubiquitous geomorphic features that occur throughout the world at mountain fronts as a result of erosion and deposition on a geologic time scale, regardless of climate. They are more prominent in semi- and arid climates because of the lack of vegetative cover that masks their characteristic fan shapes in humid areas. From both engineering and geologic viewpoints, alluvial fans present particular fluvial and sedimentation hazards in semi- and arid regions because episodic rainfall-runoff events can result in debris, mud, and fluvial flows through complex and, in some cases, migratory channel systems. Further, in semi- and arid climates alluvial fans often end in terminal or playa lakes. Given the uniform topography of playa lakes, these features often present ideal locations for facilities such as airports; however, regardless of the engineering advantages of the topography, the episodic and often long-term flooding of these lakes attracts migratory birds.

The purpose of this volume is to summarize the current state-of-the-art, from an engineering viewpoint, in the identification and mitigation of flood hazard on alluvial fans; however, to accomplish this, a fundamental understanding of geology is required. No claim is made by the authors that this book is comprehensive because of the breadth and depth of the subject worldwide; however, it is hoped that it will prove valuable in advancing the state-of-the-art knowledge.

The authors who participated in this effort are a mixture of academics and practicing engineers and geologists. Everyone involved took on this task because they wanted to provide what insight they could to understanding the issues and the approaches to this critical problem. This volume is the work of many of the best people dealing with the issue. All of the authors have worked with each other for many years, and all of them worked very hard in making this the best volume possible. We hope that the readers learn, enjoy, and appreciate the effort that went into preparing this small, but critical contribution to our understanding of the environment in which we live.

January 2010

Richard H. French, Professor
Department of Civil and Environmental Engineering
University of Texas at San Antonio
San Antonio, Texas

Julianne J. Miller, Research Hydrologist
Division of Hydrologic Sciences
Desert Research Institute
Las Vegas, Nevada

Contents

Chapter 1

Introduction

Richard H. French

Department of Civil and Environmental Engineering
University of Texas at San Antonio
6900 N Loop 1604 West, San Antonio, Texas 78249
Richard.French@utsa.edu

Alluvial fans are ubiquitous geomorphologic features that occur through-out the world, regardless of climate, at the front of mountains as the result of erosion and deposition. They are more prominent in semi- and arid climates simply because of the lack of vegetative cover that masks their fan shapes in more humid areas. From both engineering and geo-logic viewpoints, alluvial fans present particular fluvial and sedimenta-tion hazards in semi- and arid regions because episodic rainfall-runoff events can result in debris, mud, and fluvial flows through complex and, in some cases, migratory channel systems. Further, in semi- and arid climates alluvial fans often end in terminal or playa lakes. Given the uniform topography of playa lakes, these features often present ideal locations for facilities such as airports; however, regardless of the engi-neering advantages of the topography, the episodic and often long-term flooding of these lakes attracts migratory birds. The presence of migra-tory birds on playa lakes present hazards to aircraft — migratory birds are incompatible with aircraft operations. In other parts of the world such as Jordan, water on playa lakes can provide water for nomadic herders and their livestock.

This chapter is intended to introduce the reader to the funda-mental concepts with subsequent chapters covering technical details and presenting a variety of case studies. The book is primarily intended for civil engineers rather than geologists; however, as will become clear, in evaluating hazard at any given location, there should be close profes-sional interaction between the engineering and geosciences communities. The chapter concludes with an introduction to the rest of the volume.

1.1   Introduction

An alluvial fan is a "triangular or fan shaped deposit of boulders, gravel, sand, and fine sediment at the base of desert mountain slopes deposited by intermittent streams as they debouch onto the valley floor," Stone [1967]. The definition by Stone [1967] does not take into account that alluvial fans also occur in recently glaciated areas of Poland, Rachocki [1981]; the eastern and southeastern United States, Anstey [1965] and Schumm *et al.* [1996], the Rocky Mountains, *e.g.*, Telluride, Colorado; and along the Pacific Coast, *e.g.*, La Conchita, California. However, the definition given above is a reasonable generic definition to begin this volume since its focus is alluvial fans and their terminuses in semi- and arid environments. Figure 1.1 depicts a classical alluvial fan in both plan and profile.

Humans have an innate curiosity about the environment in which they live and a desire to understand the processes that formed and continue to modify their environment; however, there are at least four reasons why alluvial fans and playas (terminal lakes) are of interest to the engineering and scientific communities. First, and perhaps the primary reason, is that some of the fastest developing areas of the United States, and other portions of the world, are located in semi- and arid environments; and, in particular, the southwestern United States. In the southwestern United States alluvial fans and bajadas occupy approximately 31.3 percent of the area [Anstey, 1965].

Major urban and suburban areas such as Los Angeles, San Diego, Coachella Valley, San Bernardino County, San Diego County (all California); Tucson and Phoenix, Arizona; Salt Lake City, Utah; Las Vegas, Nevada, and other locations have in the last 60 years grown from lightly populated areas into important industrial, financial, and recreation centers. Portions, and in some cases, all of these urban and suburban areas, are located on alluvial fans and in some cases in proximity to terminal lakes, *e.g.*, Salt Lake City. As development has taken place, these areas have experienced devastating fluvial floods, mud flows, debris flows, and landslides that have resulted in the loss of life and significant property damage. The anecdotal record is replete with examples of clear-water and sedimentation hazards associated with development in semi- and arid environments [McPhee, 1989]; and as development progresses the potential for damage becomes greater.

PLAN

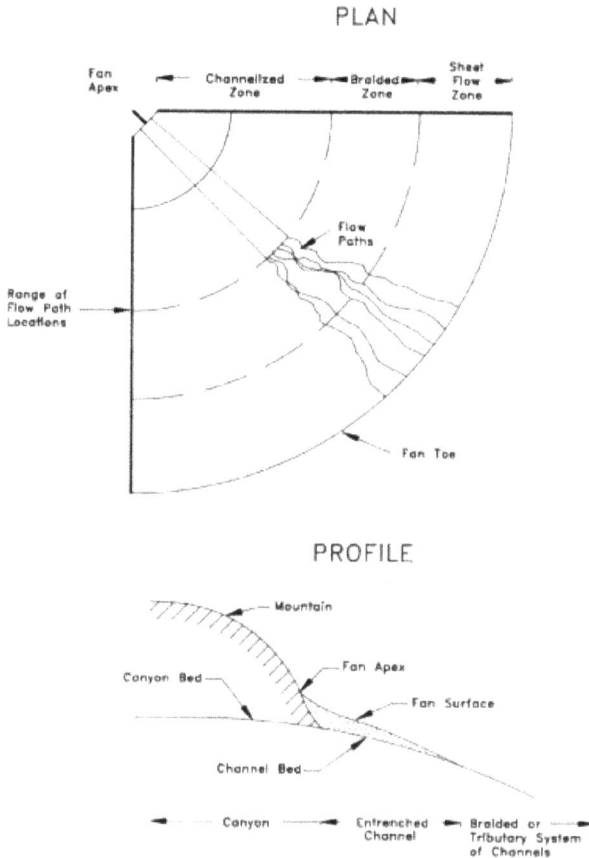

PROFILE

Figure 1.1 Plan and profile view of an idealized alluvial fan with the zones of flow suggested by Anderson and Nichols (1981) indicated.

1.2 Alluvial Fan Hazards

Whereas Stone's [1967] definition of an alluvial fan is a sufficient generic definition, there are definitions that are more relevant from an engineering viewpoint. For example, the U.S. Federal Emergency Management Agency's (FEMA) definition of an alluvial fan [Federal Register, 1989] is

> "Alluvial fans are geomorphic features characterized by cone-or fan shaped deposits of boulders, gravel, sand, and fine sediments

that have been eroded from mountain watersheds, and then deposited on the adjacent valley floor .... Flooding that occurs on active alluvial fans is characterized by fast-moving debris and sediment laden shallow flows. The paths followed by these flows are prone to lateral migration and sudden relocation to other portions of the fan. In addition, these fast moving flows present hazards associated with erosion, debris flow, and sediment transport."

The FEMA definition itemizes the hydraulic processes expected to occur on a generic, regulatory alluvial fan from an engineering viewpoint; and this definition makes clear that hazards on alluvial fans are due to a wide range of hydraulic processes that involve sediment movement and transport; and many of these processes are not yet well understood. Finally, Schumm *et al.* [1996] in a study of alluvial fan flooding for the U.S. National Research Council presented the following definition:

"Alluvial fan flooding is a type of flood hazard that occurs only on alluvial fans. It is characterized by flow path uncertainty so great that this uncertainty cannot be set aside in realistic assessments of flood risk or in the reliable mitigation of the hazard. An alluvial fan flooding hazard is indicated by three related criteria: (a) flow path uncertainty below the hydrographic apex; (b) abrupt deposition and ensuing erosion of sediment as a stream or debris flow loses it competence to carry material eroded from a steeper, upstream source area; and (c) an environment where the combination of sediment availability, slope, and topography creates an ultra-hazardous condition for which elevation on fill will not reliably mitigate the risk."

Fatalities and property damage are, unfortunately, a dependable way to quantify the magnitude of the losses experienced; however, in many cases the loss of irreplaceable family possessions and disruption of the community life is also considerable. Below, three instances are used to illustrate these issues from the viewpoint of fluvial or near-fluvial events:

(1) On September 14, 1974 a flash flood swept through El Dorado Canyon on the Nevada side of Lake Mojave, a reservoir on the Colorado River, killing at least 9 people; destroying 5 mobile homes; 38 vehicles; 19 boat trailers; obliterating a restaurant; and destroying half the extensive boat docking facilities on the lake. The peak flow from the $23\,\text{mi}^2$ ($59\,\text{km}^2$) watershed was estimated to be $77,602\,\text{ft}^3/\text{s}$ ($2,200\,\text{m}^3/\text{s}$) [Glancy and Harmsen, 1975].

(2) On July 26, 1981 a distant thunderstorm resulted in a small flash flood in Tanque Verde Creek, a popular summer recreation area near Tucson, Arizona sweeping 8 people to their deaths over Tanque Verde Falls [Hjalmarson, 1984]. Because of the fatalities, this small flood received a great deal of media attention although much larger floods occurring one day earlier and four days later received no media attention. As noted by Imhoff and Shanahan [1980], flooding in semi- and arid environments is less related to the absolute magnitude of the flood than is the case with the impact of flooding in perennial rivers and more related to the quickness and ferocity of the event.

(3) In 1983, there was landslide-induced flooding at Ophir Creek, Nevada [Glancy and Bell, 2000]. Ophir Creek is a small elongated watershed with an area of approximately $4.5 \, \text{mi}^2$ ($11.7 \, \text{km}^2$) terminating in an alluvial fan. The total sediment deposited during this event was approximately 450 ac-ft ($555{,}000 \, \text{m}^3$), and the flood surge was estimated to have a peak flow of approximately $50{,}000 \, \text{ft}^3/\text{s}$ ($1{,}400 \, \text{m}^3/\text{s}$). Structures outside the 100-year estimated regulatory floodplain were destroyed by the 30 ft (9 m) high, 100 ft (30 m) wide wall of boulders, mud, trees, and water, and one life was lost. Was this an extreme event? Common sense, given the magnitude of these estimates, suggests one answer, while the historical record suggests a different answer. From the historical record, Glancy and Bell [2000] discovered that significant flooding had taken place on the Ophir Creek alluvial fan in 1874, 1875, 1890, 1907, 1937, 1943, 1950, and 1963. It would appear that this single event confirms the old adage that those who fail to learn from history are destined to repeat it, and this was only a single event that took place along a critical transportation alignment and in proximity to two major urban areas.

Fluvial flooding on alluvial fans is not the only hazard issue; that is, mud and debris flows, which are non-Newtonian, also present serious and significant hazards in some areas. Below, two instances are used to illustrate these issues from the viewpoint of non-Newtonian flow events:

(1) Chawner [1935] reported the following data regarding the transport of boulders during flooding that occurred in 1934 at Montrose, a suburb northeast of Los Angeles, California. "The weights and distances from the canyon mouth of the greatest boulders were: 32 tons at 7,000 ft ($2.8 \times 10^5 \, \text{N}$ at 2,100 m); 23 tons at 7,900 ft ($9.8 \times 10^4 \, \text{N}$ at 2,400 m); 11 tons at 8,400 ft ($2.8 \times 10^5 \, \text{N}$ at 2,600 m); 5.5 tons at 9,400 ft ($4.9 \times 10^4 \, \text{N}$ at 2,900 m); 5 tons at 9,800 ft ($4.4 \times 10^4 \, \text{N}$ at 3,000 m);

1.6 tons at 9,600 ft (1.4×10^4 N at 2,900 m); and 1.2 tons at 11,800 ft (1.1×10^4 N at 3,600 m)."

Note in the above quotation the original U.S. customary units have been converted to SI units shown in parentheses. Chawner [1935] continued to note that during this single event approximately 700,000 yd^3 (535,000 m^3) of sediment were deposited in the study area. Given the volume, it suggests that 2.5 in (0.064 m) of material was removed from the contributing watershed; and given the size and weight of material moved, it is quite clear that this was not a Newtonian fluvial flow but a non-Newtonian debris flow.

(2) La Conchita is a small scenic beach community located approximately mid-way between Ventura and Santa Barbara, California. Beginning as far back as the 1800's, there is a historical record of landslides and mud/debris flow in this area [Hemphill, 2001]. The Southern Pacific Railroad laid tracks between what is now La Conchita and the Pacific Railroad in 1887. In 1889, sections of the track were buried twice by landslides (unlikely) or mud/debris flows (likely). At this point, the Southern Pacific Railroad bulldozed or hydraulically excavated the toe of the alluvial fan to protect the railroad and this facilitated the development of the community. In 1924, the La Conchita del Mar subdivision consisting of 330 lots was established. The cutting of the toe of the fan likely exacerbated the modern hazard in this community, but certainly was not the primary cause of the modern hazard as history demonstrates. In January 2005, a "landslide" destroyed or seriously damaged 36 structures and resulted in 10 fatalities [Jibson, 2005]. The word landslide is in quotations because the hazard was either a mud or debris flow rather than a landslide — note there is a difference between the engineering and geologic communities regarding the classification of flows. Definitive information regarding the engineering definitions of fluvial, mud, and debris flows are provided in Garcia *et al.* [2008]. Figure 1.2 depicts the results of this devastating event from the air. With regard to this photograph, note that in the background is the Pacific Ocean and Rincon Beach, the Pacific Coast Highway, the railroad alignment, and then the community. On the right side of the picture, note the structure rotated off its foundation.

In addition to the specific situations presented above, Schumm *et al.* [1996] provided an a list of alluvial fan flood characteristics and

Figure 1.2   Aerial view of the 2005 La Conchita mudslide area. For scale, the slide measures 0.25 miles (0.4 km) across (Source: Unknown).

hazards derived from examining published and unpublished reports for 23 events primarily in the Southwestern United States, but also including events in Australia and Israel. Included in this list are:

(1) High flow velocities;
(2) Flash floods and possibly translatory waves;
(3) Sheetflow and/or sheetflooding;
(4) Distributary flow;
(5) Unstable channel boundaries including bed and bank erosion and remobilization of deposited sediment;
(6) Stable channel boundaries;
(7) Movement of flow paths;
(8) Stable flow paths;
(9) Debris flow (including hyperconcentrated and mud flows); and,
(10) Alluviation.

In discussing flood hazard on alluvial fans, it should be apparent, as is discussed in Chapter 2, consideration must be given to two distinct time

Table 1.1   Estimated age and vertical accretion rate of selected alluvial fans.

Fan location	Estimated fan age (yrs)	Average vertical accretion rate (cm/1000 yrs)	Source
Arroyo Ciervo Fan San Joaquin Valley, CA	600,000	36–46	Bull (1964)
Milner Creek Fan, White Mtns., CA	700,000	7.6–15	Beaty (1970)
Fan near Frenchman Flat, Nevada Test Site, NV	7,000,000	7.1	French and Lombardo (1984)

scales involved. Fans develop on a geologic time scale; that is, over tens if not 100's of thousands of years. Note, on a geologic time scale alluvial fans are depositional features; and in Table 1.1, the estimated age and estimated average rate of vertical accretion of several alluvial fans are summarized. This table suggests that on a geologic time scale the average vertical rate of accretion is very slow; that is, alluvial fans have developed their characteristic shape over millennia rather than years. It should be emphasized that only average rates of vertical accretion for the entire fan surface can be estimated. In contrast, Beaty [1963, 1970 and 1974] concluded that no more than 10–15 percent of the material composing the alluvial fans of the White Mountains, California and Nevada is the result of normal fluvial depositional processes and 85–90 percent of the deposition on these fans is the result of successive and overlapping debris flows. Bell and Katzer [1987] asserted that in Dixie Valley, Nevada debris flow deposition is one of the primary sediment transport mechanisms currently active and the flow quantities are commonly measured in thousands of cubic meters. Consideration of deposition by cataclysmic events would significantly modify the estimated average vertical accretion rates summarized in Table 1.1.

Hidden within the geologic time scale is the engineering or human time scale which is measured in years or decades rather than millennia. For example, in the U.S. flood hazard mitigation structures, generally, must be designed to mitigate the runoff event with a return period of 100-years and in some cases 500-years. Relative to the age of alluvial fans, the lifespan of anthropogenic engineered works is insignificant; and this becomes an important engineering consideration, as will be discussed in subsequent chapters.

The primary consideration is that on a geologic time scale, flow paths across the fan surface must be erratic and unstable; otherwise, the characteristic fan shape could not be developed. Further, on a geologic time scale alluvial fans are aggradational features. However, on an engineering time scale, flow paths across a fan surface may be stable if they are not changed by development. This critical consideration has been widely discussed [*e.g.*, Schumm *et al.*, 1996, French *et al.*, 1993, French and Keaton, 1992, and Baker *et al.*, 1990] because it has an impact on hazard analysis and the design of hazard mitigation structures. Further, on an engineering time scale alluvial fans may not be aggradational, but erosional features. That is, from an engineering viewpoint, is an alluvial fan active or not? Table 1.2 provides a summary of characteristics on alluvial fans to determine whether the fan is active or not. With regard to Table 1.2, note the FEMA regulatory alluvial and active alluvial fans have much in common and similarly distributary flow systems and inactive alluvial fans have much in common. In Table 1.3, the hydraulic processes to be expected on these landforms are summarized. Finally, with regard to time scales, it is worth noting that some modern engineered structures have design lives that approach the

Table 1.2   Expected hydraulic processes on alluvial fan and distributary flow landforms (FCDMC, 1992; French *et al.*, 1993).

Active alluvial fan	Distributary flow system	FEMA alluvial fan	Inactive alluvial fan
Abandoned or discontinuous channels	Discontinuous channels	Continuous channels	Continuous channels
Channel capacity decreases downstream	No definite trend in channel capacity	Cumulative capacity constant downfan	Channel capacity increases downstream
Channel flow changes to sheetflow	Channel and sheetflow	Channelized flow (no overbank or sheetflow)	Channnelized flow (overbank flow possible)
Debris flow possible	Minor or no debris flow	Debris flow important	No debris flow
Frequent channel movement	Rare channel movement	Unpredictable channel location	Stable channels
Low channel capacity	Variable channel capacity	Channel capacity equals flow rate	High channel capacity
No calcrete (caliche)	Calcrete (caliche) horizons possible	No calcrete (caliche) considered	Calcrete (caliche) horizons

Table 1.3   Expected hydraulic processes on alluvial fan and distributary flow landforms (FCDMC, 1992; French *et al.*, 1993).

Active alluvial fan	Distributary flow system	FEMA alluvial fan	Inactive alluvial fan
No or buried varnish	Varnished surfaces possible	No or buried varnish	Varnished surfaces possible
No reddening of soils	Minor reddening of soils	No reddening of soils	Surface reddening of soils
Overall deposition	Local erosion and deposition influences flow	Overall deposition	Overall erosion
Radiating channel pattern changes to sheetflow area	Radiating channels to tributary	Single or multiple channels	Tributary drainage pattern
Slope decreases downstream	Slope increases at apex	Slope not a factor above bifurcation	Slope variable
Stream capture or avulsions?	Channel movement by stream capture	Channel movement by avulsions	No channel movement
Uniform topography (low crenulation index)	Medium to low topographic relief (medium to low crenulation index)	Uniform topography (low crenulation index)	Topographic relief (high crenulation index)
Uniform vegetation in floodplain	Diverse vegetative community	Uniform vegetation in floodplain	Diverse vegetative community
Variable channel geometry	Variable channel geometry	Regular channel geometry	Regular channel geometry
Weak soil development	Variable soil development	Weak soil development	Strong soil development

geologic time scale. For example, high level radioactive waste management sites have design lives of 10,000 years. Over that length of time, climate change and other geologic processes may be critical concerns. Although it is always important for engineers and geologists to co-operate on projects located on alluvial fans [Schick, 1974; French and Keaton, 1992; and Keaton *et al.*, 1990] when the project involved requires a design life that approaches a geologic time scale, this co-operation becomes critical and a subsequent chapter deals with this issue.

It should be recognized that alluvial fans are complex geomorphologic features that respond not only on an engineering time scale but on a geologic time scale. Further, it rare for a singular alluvial fan to occur; that is, in

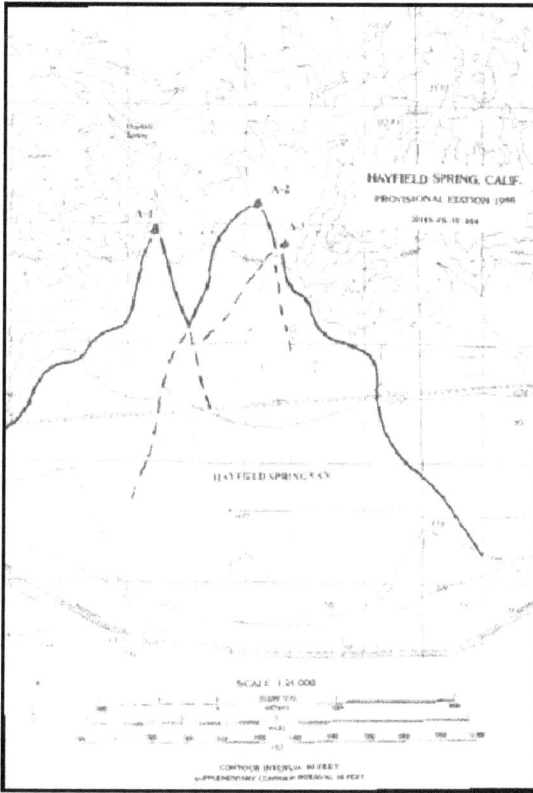

Figure 1.3   Alluvial fan surfaces above the Colorado River Aqueduct near Haywood Springs, California. There are multiple alluvial fans and training dikes protect the Aqueduct, by directing flood flows to the limited conveyance structures at the Aqueduct.

most cases hazard identification and mitigation must take into account that what were singular alluvial fans have coalesced to form a complex geomorphologic landform (Figure 1.3).

1.3   Playa Lakes

Playa or terminal lakes are essentially flat surfaces with minimal topographic relief; and they are common in most semi- and arid environments [*e.g.*, Goudie, 1991]. Both ground and surface waters can accumulate within a terminal basin and result in flooding of the playa. The terms terminal lake, playa, and playa lake are often confused or inappropriately used. For example; the Great Salt Lake, Utah and Pyramid Lake, Nevada are terminal lakes, but they are not playa lakes, from the engineering viewpoint, since they are not dry. Therefore, from the viewpoint of this volume, they are not playa lakes. In this volume, the description provided by Cooke and Warren [1973] is appropriate; that is, "The lowest areas within enclosed desert drainage basins are often marked by almost horizontal, largely vegetation free surfaces of fine-grained sediments."

Because of their minimal topographic relief and their generally dry condition, playas have historically been used as locations for airports in semi- and arid environments; for example, Edwards Air Force Base, California; Creech Air Force Base, Nevada; and potentially a new airport for Las Vegas, Nevada. Playas also present recreational opportunities; for example, dry land sailing. However, the dry lakebeds are episodically covered with water for periods ranging from days to months. The presence of water on the playa lakebed can obviously impact aircraft operations if it covers runways, taxiways or aprons. Another implication is that many playas lie on the flyways of migratory birds and when water is present the birds are attracted to it. The presence of both aircraft and birds presents a hazard to both users of the playa, and places a number of agencies in adversarial positions. Given that both military and civilian facilities are often found on playas in the U.S. and elsewhere, in the U.S. there is a requirement to delineate the 100-year regulatory floodplain; however, the pertinent regulatory agencies have provided neither guidance nor a generally accepted approach to quantifying flood hazard on these features. In response to the need to define the regulatory floodplain on playas, a relatively simple hydrologic model was developed [French *et al.*, 2005] and is described in a subsequent chapter along with the use of remotely sensed data that partially verifies the results provided by the model [French *et al.*, 2006]. From a U.S. regulatory viewpoint, it is also pertinent to note some playas may be Waters of the United States and subject to regulation under the Clean Water Act [Freeman and Rasband, 2002], which continues to be a contentious regulatory issue in the United States.

Playas can also serve as a crucial water source of water. For example, in Jordan playas (known as "Qa") have historically and currently provide episodic sources of water for nomadic Bedouins who travel with their livestock across the desert looking for water and grass [Al Qudah, 2003].

1.4   Conclusion

As will be noted later in this volume, the identification and mitigation of hydrologic and geologic hazards in semi- and arid environments is a relative new topic that is still evolving, from an engineering viewpoint (circa 1979) and from a quantitative/empirical geologic viewpoint somewhat earlier [*e.g.*, Hooke, 1965 and 1967]. It is worth observing that the problems have not changed; however, the magnitude and menace of the problem to the public and property has greatly increased as urban areas have increased in size and population in all semi- and arid environments. Has progress been made, certainly [*e.g.*, Parker *et al.*, 1998a and 1998b; O'Brien *et al.*, 1993; Cazanacli *et al.*, 2002; and French and Miller, 2003]; however, progress has been painfully slow over the past three decades. Engineering and scientific discussions continue [Schumm *et al.*, 1996] and these are serious and substantiative technical discussions.

This volume is intended to focus on the fundamentals of geologic and hydrologic processes that are particularly important in semi- and arid environments. In Chapter 2, basic hydrologic and geologic concepts are presented. While most civil engineers, worldwide, are required as part of their undergraduate studies to acquire some knowledge of geology, the curricula generally followed does not adequately prepare them for task of hazard identification and mitigation in semi- and arid environments; therefore, Chapter 2 is intended assist in filling this gap. Chapter 3 treats traditional approaches to identifying and mitigating flood hazards on alluvial fans that were put forward in the late 1970's [Magura and Wood, 1980 and Dawdy, 1979] and were and have remained controversial in both the civil engineering and geologic professional communities. This chapter presents the theory and discusses the issues associated with these approaches. Chapter 4 discusses new modeling approaches to hazard identification on alluvial fans that include modeling fluvial, mud, and debris flow hazards. Chapter 5 treats a critical engineering issue; that is, flood hazard vs. risk analysis. As discussed in Section 1.3, many alluvial fans terminate on playa lake beds. Chapter 6 presents and discusses approaches to quantifying playa lake flooding hazards. As discussed throughout this volume, most civil engineering projects can benefit from interaction between the geologic and engineering communities of practice and particularly in the identification and mitigation hazard in semi- and arid environments. Therefore, Chapter 7 discusses this need and provides examples of how this co-operation in the past has provided tangible benefits to a variety of projects. In Chapter 8, three interesting and diverse case studies are presented. The first case study discusses the Mesquite Trails project at Borrego Springs in San Diego County, California. The second case study involves the simulation of a large scale and destructive debris flow at La Conchita, California. The third study discusses work performed on Tiger Wash in Maricopa County, Arizona. In Chapter 9, some future directions are suggested in both terms of research and practice.

No single volume could hope to completely summarize the vast amount of knowledge that is available regarding the identification and mitigation of fluvial and sediment hazards in semi- and arid environments. For this volume a diverse team of authors including professionals from engineering and geologic communities and from both private sector and academia was assembled. It is anticipated that this book will summarize the current state-of-the-art and point the direction to the future.

References

Al-Qudah, K. (2003). "The influence of long-term landscape stability on flood hydrology and geomorphic evolution of the valley floor in the Northeastern Badia of Jordan." Ph.D. Dissertation, University of Nevada, Reno, Reno, NV.

Anderson-Nichols (1981). "Flood plain management tools for alluvial fans: Study findings." Anderson Nichols, Inc., Palo Alto, CA.

Anstey, R.L. (1965). "Physical characteristics of alluvial fans." Technical Report ES-20, U.S. Army Natick Laboratories, Natick, MA.

Baker, V.R., Demsey, K.A., Ely, L.L., Fuller, J.E., House, P.K., O'Connor, J.E., Onken, J.A., Pearthree, P.A. and Vincent, K.R. (1990). "Application of geological information to Arizona flood hazard assessment." *Proceedings of the International Symposium Hydraulics/Hydrology of Arid Lands*, ASCE, Reston, VA, 621–626.

Beaty, C.B. (1963). "Origin of alluvial fans, White Mountains, California and Nevada." *Annals of the Association of American Geographers*, (53), 516–535.

Beaty, C.B. (1970). "Age and estimated rate of accumulation of alluvial fan, White Mountains, California, U.S.A." *American Journal of Science*, (268), 50–77.

Beaty, C.B. (1974). "Debris flows, alluvial fans, and a revitalized catastrophism." *Z. Geomorph. N.F.*, (21), 31–51.

Bell, J.W and Katzer, T. (1987). "Surficial geology, hydrology, and late Quaternary tectonics of the IXL canyon area as related to the 1954 Dixie

Valley earthquake." Bulletin 102, *Nevada Bureau of Mines and Geology*, Reno, NV.

Bull, W.B. (1964). "Alluvial fans and near surface subsidence in western Fresno county." Professional Paper 437-A, U.S. Geological Survey, Washington, D.C.

Cazanacli, D., Paola, C. and Parker, G. (2002). "Experimental steep braided flow: Application to flooding risk on fans," *Journal of Hydraulic Engineering*, 128(3), 322–330.

Chawner, W.D. (1935). "Alluvial fan flooding: the Montrose, California flood of 1934." *Geographical Review*, 25, 255–263.

Cooke, R.U. and Warren, A. (1973). "Geomorphology in deserts," University of California Press, Berkeley, CA.

Dawdy, D.R. (1979). "Flood frequency estimates on alluvial fans," *Journal of the Hydraulics Division*, 105(HY11), 1407–1412.

FCDMC, (1992). Alluvial fan data collection and monitoring study. Prepared by: CH2M-Hill, Tempe, AZ. and R.H. French, Las Vegas, NV. For: Flood Control District of Maricopa County, Phoenix, AZ.

Federal Register (1989). 54(156).

Freeman, G.E. and Rasband, J.R. (2002). "Federal regulation of wetlands in aftermath of Supreme Court's decision in SWNACC vs. United States," *Journal of Hydraulic Engineering*, 128(9), 806–810.

French, R.H. and Lombardo, W.S. (1984). "Assessment of flood hazard at the radioactive waste management site in Area 5 of the Nevada test site" Water Resources Center, Desert Research Center, Las Vegas, NV.

French, R.H. and Keaton, J.R. (1992). "Successful interactions between hydraulic engineering and geomorphology in identifying flood hazard areas in the Southwestern United States." Proceedings of the Hydraulic Engineering Sessions at Water Forum '92. ASCE, Reston, VA, 581–586.

French, R.H., Fuller, J.E. and Waters, S. (1993). "Alluvial Fan: Proposed new process-oriented definitions for arid southwest," *Journal of Water Resource Planning and Management*, 119(5), 588–598.

French, R.H. and Miller J. (2003). Discussion of "Experimental steep braided flow: Application to flooding risk on fans," by D. Cazanacli, C. Paola, and G. Parker, *Journal of Hydraulic Engineering*, 128(3), 2002, 129(11), 920–922.

French, R.H., Miller, J.J. and Dettling, C.R. (2005). "Estimating playa lake flooding: Edwards air force base, California, USA," *Journal of Hydrology*, 306(2005), 146–160.

French, R.H., Miller, J.J., Dettling, C. and Carr, J.R. (2006). "Use of remotely sensed data to estimate the flow of water to a playa lake," *Journal of Hydrology*, 325(2006), 67–81.

Garcia, M., MacArthur, R., French, R. and Miller, J. (2008). "Sedimentation hazards," Sedimentation Engineering: Processes, Measurements, Modeling, and Practice, M.H. Garcia, ed., ASCE Manual of Practice 110, Reston, VA., 885–922.

Glancy, P.A. and Bell, J.W. (2000). "Landslide-induced flooding at Ophir Creek, Washoe County, Western Nevada, May 30, 1983." Professional Paper 1617, U.S. Geological Survey, Washington, D.C.

Glancy, P.A. and Harmsen, L. (1975). "A hydrologic assessment of the September 14, 1974 flood in El Dorado Canyon, Nevada." Professional Paper 930, U.S. Geological Survey, Washington, D.C.

Goudie, A.S. (1991). "Pans," *Progress in Physical Geology*, 15(3), 221–237.

Hemphill, J.J. (2001). "Assessing landslide hazard over a 130-year period for La Conchita, California." Association of Pacific Coast Geographers Annual Meeting

Hjalmarson, H. (1984). "Flash flood in Tanque Verdi Creek, Tucson, Arizona." *Journal of Hydraulic Engineering*, 110(12), 1841–1852.

Hooke, R. LeB. (1965). "Alluvial fans," Ph.D. Dissertation, California Institute of Technology, Pasadena, CA.

Hooke, R. LeB. (1967). "Processes on arid-region alluvial fans," *American Journal of Science*, 266, 609–629.

Imhoff, J.C. and Shanahan, E.W. (1980). "Flood plain management tools for alluvial fans: state-of-the-art report." Federal Emergency Management Agency, Washington, D.C.

Jibson, R.W. (2005). "Landslide hazards at La Conchita, California." Open-File Report 2005-1067, U.S. Geological Survey, Washington, D.C.

Keaton, J.R., Shlemon, R.J., French, R.H., and Dawdy, D.R. (1990). "Piedmont-fan flood hazard analysis from geomorphology and surface water hydrology, Hudspeth County, Texas," *Proceedings 1990 International Symposium Hydraulics/Hydrology of Arid Lands*, ASCE, New York, NY, pp. 356–360.

Magura, L.M. and Wood, D.E. (1980). "Flood hazard identification and flood plain management on alluvial fans," *Water Resources Bulletin*, 16(1), 56–62.

McPhee, J. (1989). "Los Angeles against the mountains," The Control of Nature, Straus and Giroux, New York, NY.

O'Brien, J.S., Julien, P.Y., and Fullerton, W.T. (1993). "Two dimensional water flood and mudflow simulation," *Journal of Hydraulic Engineering*, 119(2), 244–261.

Parker, G., Paola, C., Whipple, K.X. and Mohrig, D. (1998a). "Alluvial fans formed by channelized fluvial and sheetflow: Theory," *Journal of Hydraulic Engineering*, 124(10), 985–995.

Parker, G., Paola, C., Whipple, K.X. and Mohrig, D. (1998b). "Alluvial fans formed by channelized fluvial and sheetflow II: Application," *Journal of Hydraulic Engineering*, 124(10), 986–1004.

Rachocki, A. (1981). "Alluvial fans, an attempt at an empirical approach," John Wiley and Sons, Inc., New York, NY.

Schick, A.P. (1974). "Alluvial fans and desert roads — a problem in applied geomorphology," Abhandlungen der Akademie der Wissenschaften in Gottingen Mathematisch — Physiskalische Klasse, 29, 418–425.

Schumm, S.A., Baker, V.R., Bowker, M.F., Dixon, J.R., Dunne, T., Hamilton, D., Hjalmarson, H.W. and Merritts, D. (1996). Alluvial Fan Flooding, Committee on Alluvial Fan Flooding, Water Science and Technology Board, Commission on Geosciences, Environment, and Resources, National Research Council, National Academy Press, Washington, D.C.

Stone, R.O. (1967). "A desert glossary." *Earth Science Reviews*, (3), 211–268.

Chapter 2

Geologic and Hydraulic Concepts of Arid Environments

Julianne J. Miller

Desert Research Institute, Division of Hydrologic Sciences
755 East Flamingo Road, Las Vegas, Nevada 89119
Julie.Miller@dri.edu

Deserts are arid and semi-arid environments, receiving very little precipitation. The formative geologic principle of Uniformitarianism is very evident in the long, slow geologic processes that contribute to desert landscape evolution. Wind and water erosion are important mechanical weathering processes that contribute to desert landscapes. Aeolian (wind) erosion creates bedforms, such as sand dunes. Water erosion, by conveying sediment of varying concentrations, forms important geomorphic landforms, such as alluvial fans. Vegetation species are diverse, ranging from grasses and shrubs to woodlands, and, along with burrowing species, cause bioturbation of desert soils. Desert pavements and indurated layers play an important role in the ability of desert soils to infiltrate water or create runoff from precipitation events.

2.1 Introduction

Deserts are, by definition, arid and semi-arid environments, receiving less than 15 cm (6 in) and between 25 to 50 cm (10 to 20 in) of precipitation per year, respectively. Evaporation rates in these environments can be between 10 to 20 times that of precipitation [Stone, 1968]. By this definition, deserts can exist in several temperature zones, ranging from hot desert climates (discussed here) to the polar climates of the Arctic and Antarctic regions.

Deserts typically have poor vegetative cover, although the vegetation is extremely diverse, ranging from grasses and shrubs to woodlands.

Vegetative type is typically a result of soil moisture, soil chemistry, and elevation, with the grasses and shrubs found in the lower elevations and the woodlands in the mountainous regions. Vegetation plays an important role in soil stabilization, and thus the effects of wind and water erosion on a geomorphic surface.

2.1.1 *Desert landscape formation*

Wind and water are both mechanical weathering processes that are central to the evolution of the formation of desert landscapes (Figure 2.1). Wind, or aeolian, erosion of a surface is dependant upon the particle size available for transport. Silts and clay-sized particles are transported by suspension, or entrainment, within the wind. Sands typically travel by saltation, literally bouncing across a surface, whereas coarser materials tend to roll or slide across a surface [Ritter, 1978].

Geomorphic features formed in hot desert environments by aeolian transport processes are found in a large range of sizes. Common examples of this are sand bedform deposits, ranging from tiny ripples to sand dunes and finally to *draas*, giant features of the North African deserts [Ritter, 1978]. Sand dunes, an intermediate size deposit, are the most common aeolian depositional form.

Figure 2.1 Desert landscape formed by water and wind erosion.

Figure 2.2   Alluvial fans in Death Valley National Park, California. The photo on the left shows a classic alluvial fan; the photo on the right shows coalescing alluvial fans.

Water erosion also plays an important role in forming geomorphic features within arid environments. Alluvial fans, discussed in Chapter 3, are a dominant landform created by water erosion of upstream sediment materials, transported from steep mountain slopes, and then deposited in the open valleys between mountain ranges (Figure 2.2).

2.2   Geologic Theories of Formative Processes

2.2.1   *Catastrophism*

Early Christian beliefs fueled the ancient concept of Catastrophism as the mechanism responsible for the Earth's formation, assuming that the Earth formed through a series of sporadic catastrophic events, such as floods, earthquakes, and volcanoes. Early geologists thought that these sudden and violent events formed the Earth over a very short period of time, not much longer than human history. The biblical story of Noah's flood is often cited as an example of an event supporting the theory of Catastrophism.

2.2.2   *Gradualism (Uniformitarianism)*

Gradualism, the opposite of Catastrophism, assumes that geologic change occurs slowly over long periods of time. This principle became known as Uniformitarianism, with the additional assumption that the same natural processes that occur now, have always occurred in nature. The principle of Uniformitarianism is generally stated as "the present is the key to the past."

Although the roots of Gradualism and Uniformitarianism can be traced to the 11th century, the modern philosophy began in the late 18th century with James Hutton [Hutton, 1788]. Charles Lyell [Lyell, 1830] further developed Hutton's concept that the Earth had been formed by the same slow-moving forces that occur at present day. Many geologists of that time abandoned the tenets of Catastrophism that catastrophic events, such as floods, earthquakes, and volcanoes, were important events affecting the formation of the Earth over geologic time. However, most contemporary geologists now recognize the intermittent role that catastrophic events play in geologic formative processes.

2.2.3   *Integration*

Uniformitarianism is now a basic principle of geology. However, the consensus amongst today's geologists is that the Earth's geologic formation was a slow, gradual process, affected by occasional significant natural catastrophic events. Modern geologists also accept that the rates at which geologic processes occur may not have been constant over time, as Hutton believed [Marshak, 2009]. The mass extinction of species, including the dinosaurs, at the end of the Cretaceous period (65 million years ago), is now generally thought to be the result of a combination of catastrophic events, including increased volcanism and the impact of an asteroid [Marshak, 2009]. Although these catastrophic events significantly altered the geologic history of the Earth, the Earth's formation continued as a gradual process, following the principle of Uniformitarianism.

2.3   Flow Processes

2.3.1   *Fluvial*

Flows in deserts are typically produced by high-intensity, short-duration precipitation events. These precipitation events are so intense, that the desert soils cannot adequately infiltrate the water, resulting in quick, large flow events or flash floods within ephemeral channels. These flow events are of short duration, occurring over minutes or hours, with peak discharges reached sooner than floods in perennial rivers [French, 1987].

   The usual sediment transport mechanisms in water flows include suspended and bedload transport. Depending on how much unconsolidated sediment is available within a watershed will partially determine whether the water flood becomes hyperconcentrated with sediment or not.

2.3.2 *Hyperconcentrated flows*

Flood events may become hyperconcentrated with sediment. Hyperconcentrated flows, particularly debris and mudflows, are dominant hydraulic processes in the development of alluvial fans in arid and semi-arid environments [French, 1987]. These flows can be the result of rainfall, snowmelt, volcanic, high groundwater tables, and anthropegenic causes [Wan and Wang, 1994]. Flow hydraulics, cessation, and runout distances are controlled by the fluid matrix, which is dependant upon sediment concentration, size fraction, and clay content [Garcia *et al.*, 2008].

Hyperconcentrated flows, ranging from mudfloods to landslides, have been classified based on sediment concentrations, the mechanism responsible for the flow, and the rheological and kinematic behavior by a variety of researchers (Table 2.1). Definitions of the characteristics of the different flows are found below.

Mudflood

Mudfloods are hyperconcentrations of mostly non-cohesive particles, such as sand, uniformly distributed within the flow (Figure 2.3). They behave as fluids, with sediment concentrations ranging from 25 to 45 percent [Winterwerp *et al.*, 1990]. Deformation occurs with stress, and there is no yield stress.

Mudflow

Mudflows have a high concentration of silt and clay-sized particles, changing to fluid matrix to support larger clasts (Figure 2.4). They behave as highly viscous fluid masses, with sediment concentrations ranging from 45 to 55 percent. The high viscosity gives the flow considerable yield strength. Thus, mudflows can suspend large clasts over long distances, resulting in characteristic lobe-shaped deposits [Garcia *et al.*, 2008].

Debris Flow

Debris flows are less fluid than mudflows, consisting of all sizes of sediment, with mostly coarse clastic material in the frontal lobe, followed by finer grained sediments in the more fluid matrix [Takahashi, 1980] (Figure 2.5). The characteristic density and viscosity of debris flows distinguish them from other hyperconcentrated flows, with a density twice that of clear water and a significantly larger viscosity than water [French, 1987]. Because

Table 2.1   Classification of Hyperconcentrated Flows [modified from Bradley and McCutcheon, 1987 in Garcia, 2008].

	Concentration percent by weight (100% by WT = 1,000,000 ppm)									
	23	40	52	63	72	80	87	93	97	100
	Concentration percent by volume (G = 2.65)									
	10	20	30	40	50	60	70	80	90	100
Beverage and Culbertson (1964)	High	Extreme	Hyperconcentrated			Mud flow				
Costa (1984)		Water flood	Hyperconcentrated		Debris flow					
O'Brien and Julien (1985) using National Research Council (1982)		Water flood	Mud Flood	Mud Flow		Landslide				
Takahashi (1981)	Fluid flow			Debris or Grain flow				Fall, Landslide, Creep, Sturzstrom, Pyroclastic flow		
Fan and Dou (1980)	Sediment Laden		Hyperconcentrated flow			Debris or Mud flow				
Pierson and Costa (1984)	STREAMFLOW Normal: Hyperconcentrated		SLURRY FLOW (Debris torrent), Debris mud flow, Solifluction					GRANULAR FLOW Sturzstrom, Debris Avalanche, Earthflow, Soil creep		

Figure 2.3   Mudflood contained within a constructed channel [Julien and Leon, 2000].

Figure 2.4   Frontal lobe-shaped deposit of a mudflow [O'Brien *et al.*, 1993].

Figure 2.5   A debris flow deposit within a channel on an alluvial fan in the Mojave National Preserve, California.

of the large viscosity, debris flows are considered to be laminar flows, with little to no mixing between the flow layers.

Landslide

Landslides have very low water content, with high sediment concentrations [Bagnold, 1956]. Gravity is the primary force for a landslide to occur, although slope stability must be affected, and a triggering factor is needed (Figure 2.6). Natural causes of landslides include groundwater saturation, loss of vegetation, degradation of soil structure, erosion of the slope toe, earthquakes, and volcanoes. Wildfires and deforestation also aid to destabilize slopes.

2.4   Soils

Desert soils are slowly formed over long periods of time, and if preserved over time, are found as very thin profiles on stable surfaces. In general, the weathering processes form a surface that is susceptible to erosion, underlain by very resistive subsurface, which may eventually become exposed if the overlying surfaces are eroded.

Figure 2.6 La Conchita, California landslide (USGS).

2.4.1 *Soil formation in arid environments*

Much of the pedogenic material that forms desert soils consists of aeolian deposited dust [Reheis *et al.*, 1995]. The incorporation of this fine material into the surficial soil horizons provides much of the water-holding capacity of desert soils [McDonald *et al.*, 2003].

Desert soils structure consist of an upper A horizon, that is generally very thin and weak, but much more permeable than subsequent underlying horizons. If an armoring layer, such as desert pavement, has formed on the surface, it is common for a vesicle-rich Av horizon to form just beneath that layer. The underlying B and K horizons are generally less

Figure 2.7   Desert pavement on an alluvial fan surface at the Desert National Wildlife Refuge, Nevada.

permeable than the upper A horizon. These lower horizons may become so cemented that they become impermeable indurated layers, restricting infiltration.

2.4.2   *Desert pavement*

Armored surfaces covered by a layer of gravel that acts to protect the underlying horizons are common in hot desert environments. This armoring is typically referred to as desert pavement (Figure 2.7). The pavement is usually no more than one layer of gravel thick, overlying a developed soil. The upper horizon of the underlying soil usually consists of silt and clay-sized particles, and contains little to no gravel. If the sediment is carbonate rich, this layer may contain vesicles formed from off-gassing of dissolved calcium carbonate, and is referred to as a "vesicular A horizon" or an Av horizon (Figure 2.8). Soil peds of an Av horizon with visible vesicles are shown in Figure 2.9.

There are two main theories as to how desert pavements form in arid environments [Ritter, 1978]. The first is that wind and water erode the fine

Figure 2.8   An underlying Av horizon and indurated layer beneath a desert pavement surface on the Desert National Wildlife Refuge, Nevada.

Figure 2.9   Soil peds of an Av horizon, with visible vesicles.

materials from the alluvial deposits, leaving a lag of gravel; however, there is little evidence to support this theory. The second, more widely accepted theory is that clasts are lifted through the alluvium by wetting and drying cycles. As the soils expand upon wetting, the clasts move upward towards the surface, with the fines filling in below them.

2.4.3   *Indurated soil layers*

Indurated soil layers can form when the overlying surface is more permeable than the underlying surface, a common feature of desert soils. In a soil profile, the underlying B and K horizons are generally less permeable than the upper A horizon. As infiltrating water slows in the less permeable layer, minerals are deposited within the pore space, further restricting flow, and making the layer even less permeable. These lower horizons may become so cemented that they form impermeable indurated layers, restricting infiltration. If clay is present, an argillic layer can form in the B horizon. In a carbonate-rich environment, calcrete, or caliche, layers will form as a K horizon in the soil profile.

In a K horizon (calcrete), the pore space is partially or completely filled with carbonate, to the point where it forms a laterally continuous layer. If the pore space is completely filled, the layer becomes impermeable, and infiltrated water ponds on the surface of the carbonate layer at depth. Typically, K horizons form at depth, although where in the soil profile they form can vary. When erosion of the overlying surfaces occurs, these indurated surface form resistant caps over older surfaces (Figure 2.10).

2.4.4   *Vegetation and biologic role in soil development*

An integral relationship exists between soil development and biologic influence in arid and semi-arid environments [Monger and Bestelmeyer, 2006]. Bioturbation, by either vegetation roots or by burrowing vertebrate and invertebrate species, aids in soil development. Burrows are more common in undercanopy areas, where burrowing species contribute to plant mounds by bringing clasts and finer materials to the surface. Bioturbation also creates macropores that locally increases infiltration capacity of the soils [Shafer *et al.*, 2007]. Vegetation continues to grow with the increased water availability, providing a food source and shelter for burrowing species.

Figure 2.10   An exposed indurated layer (calcrete) forms a resistant cap on older surfaces along a modern incised channel, Desert National Wildlife Refuge, Nevada.

Figure 2.11   Vegetation at four surfaces of different age in the northern Mojave Desert. Clockwise from the upper left, sites are from youngest to oldest.

2.5 Runoff, Infiltration Potential, and Transmission Losses

2.5.1 *Runoff and infiltration potential*

Runoff and infiltration characteristics of desert soils vary substantially across surfaces of different ages and compositions. Alluvial landforms typically have surfaces covered by discontinuous lags of gravels, cobbles, and boulders that intermittently form moderately- to strongly-developed desert pavements. On older surfaces, the original surface may have been eroded, exposing once underlying indurated surfaces. As the soil geomorphic surfaces become older and finer particles are added through dust deposition and pedogenic processes, mean particle diameter decreases and standard deviation increases. These processes control both soil structural development and ultimately the soil's infiltration capacity and runoff potential.

2.5.2 *Channel transmission losses*

Transmission losses along ephemeral channels are an important, yet poorly understood, aspect of the rainfall-runoff process. Losses occur as flow infiltrates channel bed, banks, and floodplains. Estimating transmission losses in arid environments is difficult because of the variability of surficial geomorphic characteristics and infiltration capacities of soils and near surface low permeability geologic layers (e.g., calcrete). Transmission losses in ephemeral channels are nonlinear functions of discharge and time [Lane, 1972], and vary spatially along the channel reach and with soil antecedent moisture conditions [Sharma and Murthy, 1994].

   In studies performed at the U.S. Department of Energy/National Nuclear Security Administration's Nevada Test Site (NTS), transmission losses ranging between 40 and 70 percent of the flow were measured in channels of 1,000 to 2,000 m (3,300 to 6,600 ft) [Miller *et al.*, 2003; Mizell *et al.*, 2005]. The greatest losses were measured in the reaches of the channels with the youngest soils, characterized by unconsolidated sands, no calcrete, and no indication of long term stability. Losses were less in reaches where geomorphic surface ages were older and calcrete was present. The variation in volume of transmission losses in intermediate reaches demonstrates the important roles that desert geomorphic surface types and ages, vegetative cover and types, subsurface indurated layers (calcrete), channel slopes, and soil hydraulic properties play in runoff response.

References

Bagnold, R.A. (1956). "Flow of cohesionless grains in fluids." *Transactions of the Royal Society of London*, 249(964), 235–297.

Beverage, J.P. and Culbertson, J.K. (1964). "Hyperconcentrations of suspended sediment." *Journal of Hydraulic Engineering*, 126, 158–159.

Bradley, J.B. and McCutcheon, S.C. (1987). "Influence of large suspended-sediment concentrations in rivers." in *Sediment Transport in Gravel-bed Rivers*, Thorne, C.R., Bathurst, J.C. and Hey, R.D. (eds.), Wiley, New York.

Costa, J.E. (1984). "Chapter 9: Physical geomorphology of debris-flow." Development and applications of geomorphology, Costa, J.E. and Fleisher, P.J. (eds.), Springer, NY.

Fan, J. and Dou, G. (1980). "Sediment transport mechanics." Proceedings of International Symposium on River Sedimentation, Guanghua Press, Beijing, China, pp. 1167–1177.

French, R.H. (1987). Hydraulic processes on alluvial fans. Elsevier. New York, NY., 243p.

Garcia, M.H. (ed.). (2008). *Sedimentation Engineering: Processes, Measurements, Modeling, and Practice,* ASCE Manuals and Reports on Engineering Practice No. 110, American Society of Civil Engineers, Reston, Virginia.

Garcia, M.H., MacArthur, R.C., French, R.H. and Miller, J.J. (2008). "Sedimentation Hazards." in *Sedimentation Engineering: Processes, Measurements, Modeling, and Practice,* Garcia, M.H. (ed.), ASCE Manuals and Reports on Engineering Practice No. 110, American Society of Civil Engineers, Reston, Virginia.

Hutton, J. (1788). Theory of the Earth. In *Transactions of the Royal Society of Edinburgh*, 1(2), 209–304.

Julien, P.Y. and Leon, C.A. (2000). "Mud floods, mudflows and debris flows: classification, rheology and structural design." International Seminar on the Debris-flow Disaster of December 1999, Institute of Fluid Mechanics, University of Central Venezuela, Caracas, Venezuela (CD ROM).

Lane, L.J. (1972). "A proposed model for flood routing in abstracting ephemeral channels." *Hydrology and Water Resources in Arizona and the Southwest*, 2(2), 439 453.

Lyell, C. (1830). The Principles of Geology. Murray, London. Vol. 2.

Marshak, S. (2009). Essentials of Geology. W.W. Norton & Company, Inc. New York, NY. Paginated by section.

McDonald, E.V., McFadden, L.D. and Wells, S.G. (2003). Regional response of alluvial fans to the Pleistocene-Holocene transition and alluvial fan deposition in the Providence Mountains. In Enzel, Y., Wells, S.G., Lancaster, N. (eds.), *Paleoenvironments and Paleohydrology of the Mojave and southern Great Basin deserts*. Geological Society of America Special Paper 368, pp. 189–205.

Miller, J.J., French, R.H., Young, M.H. and Mizell S.A. (2003). *Effect of Soil Condition on Channel Transmission Loss during Ephemeral Flow Events.* American Society of Civil Engineers, Environmental and Water Resources Institute, World Water and Environmental Congress, Conference Proceedings, June 23–27, 2003, Philadelphia, PA.

Mizell, S.A., Miller, J.J. and French, R.H. (2005). *Effect of Soil Condition on Channel Transmission Loss during Ephemeral Flow Events — Part 2.* American Society of Civil Engineers, Environmental and Water Resources Institute, World Water and Environmental Congress, Conference Proceedings, May, 2005, Anchorago, AK.

Monger, H.C. and Bestelmeyer, B.T. (2006). "The soil-geomorphic template and biotic change in arid and semi-arid ecosystems." *Journal of Arid Environments*, 65, 207–218.

National Research Council. (1982). "*Selecting a Methodology for Delineating Mudslide Hazard Areas for National Flood Insurance Program*," National Academy of Sciences Report by the Advisory Board on the Build Environment, Washington, D.C.

O'Brien, J.S. and Julien, P.Y. (1985). "Physical properties and mechanics of hyper-concentrated sediment flows." In *Proceedings of the ASCE Specialty Conference on Delineation of Landslides, Flash Flood and Debris-Flow Hazards*, Bowles, D. (ed.), Utah Water Research Laboratory, Utah State University, Logan, Utah, 260–279.

O'Brien, J.S., Julien, P.Y. and Fullerton, W.T. (1993). "Two dimensional water flood and mudflow simulation." *Journal of Hydraulic Engineering, ASCE*, 119(2), 244–261.

Pierson, T.C. and Costa, J.E. (1984). "A rheological classification of subaerial sediment-water flows." 97th *Annual Meeting, Geological Society of America*, 16(6), p. 623.

Reheis, M.C., Goodmacher, J.C., Harden, J.W., McFadden, L.D., Rockwell, T.K., Shroba, R.R., Sowers, J.M. and Taylor, E.M. (1995). "Quaternary soils and dust deposition in southern Nevada and California." *Geological Society of America Bulletin*, 107, 1003–1022.

Ritter, D. (1978). Process Geomorphology. Wm. C. Brown Company Publishers. Dubuque, Iowa.

Shafer, D.S., Young, M.H., Zitzer, S.F., Caldwell T.G. and McDonald, E.V. (2007). "Impacts of interrelated biotic and abiotic processes during the past 125,000 years of landscape evolution in the Northern Mojave Desert, Nevada, USA." *Journal of Arid Environments*, 69, 633–657.

Sharma, K.D. and Murthy, J.S.R. (1994). "Estimating transmission losses in an arid region." *Journal of Arid Environments*, 26(3), 209–219.

Stone, R. (1968). Deserts and desert landforms. In *Encyclopedia of geomorphology*, Fairbridge, R.W. (ed.), Rehinhold Book Corp., New York. p. 271–279.

Takahashi, T. (1980). "Debris flow on a prismatic open channel." *Journal of Hydraulic Engineering, ASCE* 106, (HY8), 1153–1169.

Takahashi, T. (1981) Debris flow, *Annual Review of Fluid Mechanics*, 13, 57–77.

Wan, Z. and Wang, Z. (1994). "Hyperconcentrated flows." IAHR Monograph, Balkema, Rotterdamn, The Netherlands.

Winterwerp, J.C., de Groot, M.B., Masterbergen, D.R. and Verwoert, H. (1990). "Hyperconcentrated sand-water mixture flows over flat-bed." *Journal of Hydraulic Engineering, ASCE*, 116(1), 36–54.

Chapter 3

Traditional Approaches to Flood Hazard Identification and Mitigation on Alluvial Fans

Richard H. French

Department of Civil & Environmental Engineering
University of Texas at San Antonio
6900 N Loop 1604 West, San Antonio, Texas 78249
Richard.French@utsa.edu

Julianne J. Miller

Desert Research Institute, Division of Hydrologic Sciences
755 East Flamingo Road, Las Vegas, Nevada 89119
Julie.Miller@dri.edu

Flood hazard identification and mitigation in a qualitative sense has a long qualitative history and a much shorter quantitative history beginning in the late 1970's when the U.S. Federal Emergency Management Agency (FEMA) began to provide guidance regarding the identification of regulatory flood hazard on alluvial fans. From the day the technical guidance was issued it was controversial and remains so as of the date of this volume. It is also noteworthy that the original FEMA technical guidance only took into account fluvial flood hazard and did not discuss sedimentation hazards such as debris and mud flows. Further, the guidance only discussed how to identify hazard for the development of Flood Insurance Rate Maps (FIRM's) and did not discuss appropriate analytic methods to mitigate the hazard. In this chapter, the technical and philosophic origins of the FEMA guidance are presented and critically discussed along with the many, and in some cases, valid criticisms of the approach. In Chapter 4, new approaches to the identification and mitigation flood hazard on alluvial fans are discussed.

3.1　Introduction

Previous to the 1970's, with exceptions [*e.g.*, Hooke, 1965 and 1967], studies pertaining to flood hazard on alluvial fans were, from an engineering
viewpoint, primarily qualitative rather than quantitative. The qualitative
studies, many excellent, had their point of origin in the geoscience community. This is not to suggest that the engineering community nor those
in charge of regulating development on alluvial fans were not cognizant
of the hazards on alluvial fans presented by fluvial floods and sedimentation events, such as mud and debris flows; rather it is simply to note
there were no appropriate analytic tools to address these hazards. Perhaps,
the first critical engineering contribution to develop the required analytic
engineering tools was that of Magura and Wood [1980] and followed by
Dawdy [1979]. Note, Dawdy's development preceded that of Magura and
Wood's publication but was based on the Magura and Wood [1980] publication. Dawdy's [1979] development was the fundamental basis of the Federal
Emergency Management Agency (FEMA) technical guidance for the identification of flood hazard on alluvial fans. From the beginning, this guidance
[FEMA, 1983] was controversial for a variety of reasons. First, there are the
technical assumptions that underlie the methodology, and some of the minimal data supporting these technical assumptions are suspect. Second, the
methodology was stochastically based rather than deterministic; which was
a foreign assumption to most civil engineers. That is, most engineers and
physical scientists are educated with deterministic methodologies and models rather than stochastic assumptions and models; and therefore, many
engineers were uncomfortable with a stochastic model. Third, typical of
regulatory models, the FEMA approach tacitly assumed that all alluvial
fans result from and continue to be modified by the same hydrologic and
hydraulic processes — an assumption that cannot be justified [French *et al.*,
1993 and Schumm *et al.*, 1996]. The original FEMA model has repeatedly
been criticized, in many cases this criticism has been valid; and the overall
concepts were subject to review and discussion in the National Research
Council report, Schumm *et al.* [1996] and Garcia *et al.* [2008]. The other
fundamental flaw, from an engineering viewpoint, was that while the FEMA
guidance provided guidance for alluvial fan flood hazard identification, it
provided no guidance for flood hazard mitigation. That is, if a parcel of
land is identified as being within the regulatory 100-floodplain on a FEMA
Flood Insurance Rate Map (FIRM); then to build without being subject,
in most cases, to purchasing flood insurance how does an owner remove

the property from the floodplain? The FEMA [1983] guidance is silent on this fundamental engineering design issue, although they have the final decision regarding whether a parcel can been removed from the regulatory floodplain or not. Therefore, *ad hoc* approaches were developed by local regulatory agencies, all of which were subject to FEMA decisions regarding appropriateness of approach and result since a removal of the property from the regulatory floodplain by FEMA was required. The whole regulatory process has therefore involved not only technical controversies but also socio-economic-political issues. This chapter is intended to present the basic original technical philosophy to identify flood hazard, the modifications made to design hazard mitigation, and the technical and philosophic objections that have been raised.

3.2 Background

In recognition of the special fluvial flooding problems associated with flooding on alluvial fans in semi- and arid environments, FEMA [1983] promulgated a special methodology for assessing fluvial flood hazard for insurance purposes on alluvial fans. The development of FIRM maps requires a quantitative methodology for estimating flood zone boundaries. It is pertinent to note that many State Bureaus of Geology have developed flood hazard maps for alluvial fans based on the age of deposits found; however, these maps were deemed unsatisfactory, although supportive, for flood insurance purposes.

The FEMA [1983] approach for defining the flood hazard zones on an alluvial fan explicitly or implicitly requires a number of assumptions. The assumptions will first be presented and subsequently discussed. Among the assumptions, either explicitly or tacitly, required are the following [Dawdy, 1979; Magura and Wood, 1980; and FEMA, 1983]:

(1) Channel Location: During major flood events on active alluvial fans (see Tables 1.2 and 1.3), the flow does not spread evenly over the total surface of the fan; rather it is confined to a portion of the fan surface which conveys the water from the hydrologic apex to the toe. Below the hydrologic apex (note, the hydrologic and geologic apexes are often not the same) of the fan or the zone of permanent channel entrenchment (see Figure 1.1), the channel will occur at random locations at any point on the fan surface. That is, the flow is no more likely to occur in an old flow path than it is to follow a new

one. With this assumption, the stochastic nature of the approach becomes apparent. Subsequent philosophic modifications to the original model noted that below the hydrologic apex the flow is confined to a single channel, but at some downstream point, the single channel bifurcates and below this point the flow is conveyed in multiple channels.

(2) Channel Shape: The channel that conveys a flood flow over a fan surface can be approximated as a rectangular channel.

(3) State of Flow: Flow across a fan surface occurs at a critical depth and velocity. The channel in which the flow occurs is formed by the flow itself (that is, a flow-formed channel) and adjusts its dimensions to maintain the critical flow condition.

(4) Probability of Point Flooding: The probability of a point on the fan being flooded during a particular flood event decreases from the hydrologic apex to the toe of the fan because of the widening or expansion of the fan surface in the downstream direction. That is, this widening provides a greater transverse distance over which a channel of given width can occur.

(5) Avulsions: The terminology of avulsion as used by FEMA refers to the possibility that during a major flood event the flow may suddenly abandon one channel and form a new channel. Again, this assumption, demonstrates the probabilistic nature of the approach.

(6) Hydrologic Fan Apex Discharge Frequency Distribution: A flood peak discharge frequency distribution must be available at the apex of the fan.

(7) Coalescent Fans: As noted in Chapter 1 (Figure 1.3), alluvial fans seldom occur as isolated geomorphic features; rather, they occur as a part of complex geomorphic features where, over geologic time, isolated fans have joined or coalesced with adjacent fans. In the case of coalescent fans, the probability of a point being flooded is estimated by computing the probability of flooding from each source, and then combining these probabilities for the point of interest under the assumption that the probability distributions are independent.

(8) Infiltration and Precipitation: Infiltration on the fan surface is not considered and neither is on-fan precipitation.

From the viewpoint of a civil engineer educated in deterministic science, there is an inherent hesitancy of dealing with an analytic approach to hazard identification that is stochastic in nature; however, there are also serious technical issues with the assumptions, which are discussed in the following section.

3.3   Technical Issues Regarding the Assumptions

In this section, the assumptions necessary for the "traditional" method of identifying fluvial flood hazard on alluvial fans are critically examined. As noted in Section 3.1, this approach has been subject to both valid and invalid criticism since its publication and adoption. The invalid criticism generally results from landowners whose land values are decreased by being included in the floodplain; however, there is also a body of technical criticism that is valid, as will be noted in the following.

With regard to the assumption regarding **Channel Location**: on a geologic time scale, this assumption is certainly valid; otherwise alluvial fans could not develop their characteristic fan shape; that is, over geologic time the channels providing sediment to rather symmetrical aggradational landforms must migrate across the surface of the landform. However, on an engineering time scale this may not be the case, see Tables 1.2 and 1.3. In these tables, if the characteristics are compared the FEMA alluvial fan and the active alluvial fan exhibit a commonality; however, not all alluvial fans are under current climatic and geologic conditions active. Under current climate conditions, some alluvial fans are erosional rather than depositional features. Among the factors determining whether a fan is active or not are annual average rainfall-runoff, but likely more important is how that rainfall and runoff is distributed in time. For example, Leopold [1951] and Bull [1964] presented data that emphasizes the effects of changes in the amount and seasonal timing of precipitation on erosion and deposition in semi- and arid environments on the landscape. Leopold [1951] began by asserting that although annual values of temperature and precipitation in the southwestern United States do not exhibit significant trends, annual values may hide or mask short-term patterns that may have great significance. For example, short duration, low intensity precipitation events contribute very little to erosion and sediment transport because they produce little or no runoff. However, if this type of event occurs during the summer, they may provide the moisture required for the preservation and propagation of vegetation, which stabilizes the landscape. Large precipitation events may penetrate the deeper soil layers and promote vegetative growth, but they also produce a large proportion of the runoff and the accompanying sediment transport. To support his hypothesis, Leopold [1951] examined the records of precipitation in New Mexico and determined that in the period of time following 1850 Santa Fe, New Mexico experienced a relatively low frequency of small precipitation events in both the summer (June–September) and winter (October–May) periods. During this period of time, there was also a relatively high frequency of large precipitation events. Beginning in 1896 and continuing until 1939, these trends in frequency reversed themselves. In about 1850, the southwestern United States began to experience serious erosion. Although overgrazing has been usually blamed for the erosion, Leopold's [1951] thesis is that perturbations (variability) in the relative frequency of precipitation events may have weakened the vegetation sufficiently that overgrazing was only the triggering event rather than the singular cause of erosion. Bull [1964] investigated channel trenching on alluvial fans, using the same approach as Leopold [1951], in central California; and came to essentially the same conclusion. The point is that alluvial fans are complex landforms whose current state are the result of many complex climatic, geologic, and hydrologic factors; and therefore, in characterizing and analyzing a specific situation the engineer must use a great deal of judgment and seek informative interactions with professionals from other disciplines. The importance of these interactions is demonstrated in subsequent chapters such as Chapters 7 and 8. However, having made these comments, one must also note the experimental data from physical laboratory experiments acquired and presented in Cazanacli *et al.* [2002] and discussed by French and Miller [2003]. It is pertinent to note that the web accessible video that was referenced in Cazanacli *et al.* [2002] clearly demonstrates that over time, this assumption is valid.

The assumption that at some point below the hydrologic apex the single, flow cut channel bifurcates is reasonable; however, identifying the point at which this happens has been a continuing problem, which to date has no generic solutions. Determining the likely bifurcation point is generally based on a review of topographic maps, aerial photographs, and field inspections.

With regard to the assumption regarding **Channel Shape**: while there are anecdotal observations, there are no engineering or scientific data to support this global assumption. In the field of hydraulic engineering, the assumption of rectangular channels is common in theoretical developments because of the ease of dealing with the subsequent development of concepts. It is simply a convenient assumption, and likely has a minimal effect on the results.

With regard to the assumption regarding the **State of Flow**: from a technical viewpoint, this is a tenuous assumption and supported by very little technical data and to the knowledge of the authors no actual field measurement or observations. The topographic data available does suggest that at least the upper portions of the fan could support critical flow depending on the flow rate and the width of the channel; however, the width of the flow cut channel is a serious technical issue. The first author has seen in the vicinity of Hawthorne, Nevada, a flow cut channel (approximately rectangular in shape) with channel avulsions which were the result of a major flood event, event return period unknown. The critical portion of this assumption is the adjustment of the channel width to maintain critical flow. Magura and Wood [1980] and Dawdy [1979] both asserted that alluvial fan channels stabilize at the point where a decrease in depth would result in a two-hundred fold increase of width or

$$\frac{dy}{dT} = -0.005 \tag{3.1}$$

It is pertinent to note that Dawdy's [1979] assertion resulted from the Magura and Wood [1980] paper with the differences in publication dates likely resulting from journal publication schedules. The critical observation is that no additional engineering or scientific data have been provided to validate this assumption. This assertion/assumption is a critical assumption in the traditional analysis to identify regulatory flood hazard on alluvial fans.

Combining Eq. (3.1) with the assumption of critical flow, the corresponding depth of flow is

$$y = \phi Q^{0.4} \tag{3.2}$$

where Q = flow rate, y = depth of flow, and $\phi = 0.07$ if the U.S. Customary System (USCS) of units is used and 0.09 if the SI system is used. The width of the channel is

$$T = \beta Q^{0.40} \tag{3.3}$$

where T = channel width and $\beta = 9.41$ if the USCS system is used and 12 if the SI system is used. The velocity of flow is then

$$u = \frac{Q}{Ty} \tag{3.4}$$

where u = velocity of flow. It is appropriate to note that alternative equations for these variables have been developed based on the Manning equation that do not involve the assumption of critical flow [Edwards and Thielman, 1984].

With regard to the assumption regarding the **Probability of Point Flooding**: overall this is fundamentally a reasonable assumption on active alluvial fans; however, in site specific applications [FEMA, 1983], it may not be reasonable. This assumption has been interpreted to mean that along an alluvial fan contour the flood hazard probability distribution is uniform; that is, given a contour length the probability of any point on the contour being flooded is equal. This tacit assumption has been questioned by many; for example, French [1992], Cazanacli *et al.* [2002], and French and Miller [2003]. The currently available, laboratory and field, data suggest that points on contours in the vicinity of the centerline of the fan have a higher probability of flooding than points near the fan boundaries. Finally, anthropogenic features such as roads, borrow pits, gravel pits, and power lines will affect flows on alluvial fans. Roads, in particular, can have an important affect by intercepting, collecting, and diverting flood waters [*e.g.*, Schick, 1974].

With regard to the assumption regarding **Avulsions**: there is no question that channel avulsions do occur on active alluvial fans (Tables 1.2 and 1.3). The physical modeling data, on which the Cazanacli *et al.* [2002] paper is based, along with the video of the experiments, confirm this assumption. This once again challenges the engineer to use judgment determining the type of fan of interest — active or inactive? That is, inactive alluvial fans may have stable, entrenched channels and under current climate condition may not behave in the stochastic manner of active alluvial fans. There are multiple reasons why an avulsion may occur including, but not limited to, blockage of the active channel by debris and or sediment, capture of another channel because the flow overtops the banks of the active channel, and the existence of erosion resistant sub-surface structure that results in the overtopping of the banks of the active channel. In the traditional analysis, the possibility of channel avulsions is taken into account, as discussed below, with an avulsion coefficient (A) which has the range $1.0 \leq A \leq 2.0$. If $A = 1.0$ the channel never avulses and if $A = 2.0$ the channel always avulses. Lacking other evidence A is usually assumed to have a value of 1.5.

With regard to the assumption regarding **Hydrologic Fan Apex Discharge Frequency Distribution**: this is an essential and necessary assumption; however, its implementation has controversy associated with it. Following standard U.S. regulatory and engineering practice a log-Pearson Type 3 (LP3) distribution is assumed to describe the hydrologic distribution of peak flows at the hydrologic apex of the fan [WRC, 1981]. It is appropriate to recall that the LP3 distribution involves three parameters: the log average value $(\overline{X})$, the log standard deviation (S), and the log skew coefficient (G). However, there are generally no flow gages at the apices of alluvial fans; and therefore, in practice, synthesized peak flow rates at the apex derived from rainfall-runoff models for the watershed above the apex are used. A detailed discussion of rainfall-runoff modeling is beyond the scope of this volume; however, generally a tacit assumption in performing such modeling for alluvial fans is that the return period of the precipitation and the resulting runoff are the same using a rainfall-runoff model such as HEC-1 or HEC-HMS [HEC, 1990 and 2000]. It is appropriate to observe that methods other than the deterministic modeling of the rainfall runoff process are available such as regional regression of peak flow data; for example, Thomas *et al.* [1994]. While this is a tenuous assumption, it is necessary. Then, using the rainfall-runoff modeling results, synthetic estimates of $\overline{X}$, S, and G can be computed using [WRC, 1981] the following equations:

$$G = -2.50 + 3.12 \frac{\log(Q_{0.01}/Q_{0.10})}{\log(Q_{0.10}/Q_{0.50})} \tag{3.5}$$

Eq. (3.5) is valid for $-2.0 \leq G \leq 2.5$

$$S = \frac{\log(Q_{0.01}/Q_{0.50})}{K_{0.01} - K_{0.50}} \tag{3.6}$$

and

$$\overline{X} = \log(Q_{0.50}) - K_{0.50}S \tag{3.7}$$

where $Q_{0.01}$, $Q_{0.10}$, and $Q_{0.50}$ = peak flood discharges associated with exceedance probabilities of 0.01, 0.10, and 0.50, respectively; and $K_{0.01}$ and $K_{0.50}$ = LP3 deviates associated with the exceedance probabilities of 0.01 and 0.50, respectively. While the LP3 is the regulatory standard for estimating peak flows, there is and has been controversy around this model for many years and because it is a three parameter log-based model it is sensitive to parameter estimates, particularly the skew coefficient G. While the LP3 distribution is flexible, it tacitly assumes a stationary series (consider El Nino, La Nina, and normal periods) and that the rainfall-runoff modeling has been done correctly; none of the foregoing are often the case. Many have discussed the issue of the use of the LP3 distribution in semi- and arid climates including French and Miller [2006 and 2002], Reich *et al.* [1990], and Reich and Renard [1990].

A critical issue in semi- and arid environments with using rainfall-runoff modeling to develop synthetic LP3 parameters, or in rare cases actual peak flow data, is the skew coefficient, which can have a significant influence on the results [Reich and Renard, 1990]. Detailed rainfall-runoff modeling using HEC-1 or HEC-HMS [HEC, 1990 and 2000] often results in LP3 skew coefficients that are not appropriate. Note, in Eq. (3.4), estimate of the synthetic skew coefficient is based on peak flows with return periods of 2- and 100years, widely disparate events in terms of likelihood. The errors resulting from using rainfall runoff modeling results to estimate a synthetic skew coefficient can be mitigated by using engineering judgment since in most modeling the infiltration or initial abstraction assumes that the precipitation loss rates remain constant regardless of the return period of the event. Experience shows this is not the case; that is, when the event with a 2-year return period occurs, the soils are generally dry and when the 100-year event occurs the soils are likely close to saturation. A further check on the modeling involves not engineering judgment but common sense; that is, in semi- and arid environments the flow event with a 2-year return period should have a small peak flow and reasonableness can be brought by using regional regression results [*e.g.*, Thomas *et al.*, 1994]. That is, the results from detailed analysis using a rainfall-runoff model should be in general agreement with the results from a valid regional regression model. A further complication enters from the need to consider desert soils and their ages, as discussed in Chapter 2. The WRC [1981] provided a figure for the U.S. showing generalized LP3 skew coefficients of annual maximum streamflows. If the estimated synthetic skew varies widely from the generalized skew coefficient, then the parameter and variable values used in the rainfall-runoff model should be examined.

With regard to the assumption regarding **Coalescent Alluvial Fans**: as noted above, alluvial fans seldom occur as singular features; rather, over geologic time adjacent fans join to form a complex landform. Overall, there is no criticism of this assumption, and it seems reasonable; however, to the knowledge of the author there are minimal observational data. However, coalescent fans can present challenge in the identification and mitigation of flooding hazards; that is, where does one fan end and the adjacent fan begin? In the typical case, the available topographic data provides indications of the boundary but not definitive evidence of location of that boundary, which is important in the traditional hazard identification analysis. In some cases aerial photographs may provide assistance; however, depending on the importance of the project a field inspection involving both engineers and members of the geoscience community may be necessary.

3.4 Implementation of the Assumptions

The first step in implementing the traditional method of alluvial fan flood hazard identification involves transforming the LP3 variables. Defined as:

$$m = \overline{X} - \frac{2S}{G}$$

$$\lambda = \frac{4}{G^2}$$

and

$$a = \frac{2}{GS} - 0.92$$

Then, the transformed probability distribution variables are (note, dependence on the value of the skew coefficient):
For $G \neq 0$

$$\overline{Z} = m + \frac{\lambda}{a} \tag{3.8a}$$

$$S_z^2 = \frac{\lambda}{a^2} \tag{3.9a}$$

$$G_z = \frac{2}{\sqrt{\lambda}} \tag{3.10a}$$

$$C = \left(\frac{2a}{GS}\right)^{\lambda} \exp(0.92\,m) \tag{3.11a}$$

For $G = 0$

$$\overline{Z} = \overline{X} + 0.92\,S^2 \tag{3.8b}$$

$$S_z = S \tag{3.9b}$$

$$G_z = G = 0 \tag{3.10b}$$

$$C = \exp(0.92\,\overline{X} + 0.42\,S^2) \tag{3.11b}$$

Dealing with computing these transformed variables is tedious and prone to computational errors; therefore, FEMA [1990] developed an open source computer program to perform this task.

3.4.1 *Understanding the traditional approach*

A useful approach to understanding the underlying concept on which the stochastic alluvial fan methodology is based can be derived by repeating the analogy given in FEMA [1990]. A small structure is to be built on a perfect cone. This structure will be 145 m (475.7 ft) from the apex of the cone where the circumferential distance is 900 m (2,952.8 ft).

A curmudgeon lives at the apex of the cone and his activities cause the builder a problem. That is, the curmudgeon has a collection of steel balls ranging in diameter from 3 to 18 m (9.8 to 59.1 ft) in diameter in 3 m (9.8 ft) increments, and the balls are numbered 1 through 6 with the 3 m (9.8 ft) ball numbered 1, the 6 m (19.7 ft) ball numbered 2, *etc.* Once a year, the curmudgeon rolls a die to choose a ball from the collection. If he rolls a "1," he chooses a 3 m (9.8 ft) ball; if he rolls a "2," he chooses a 2 m (6.6 ft) ball, *etc.* Having chosen a ball, he places it at the apex of the cone and releases it. Since the cone is perfect, the ball rolls down the cone taking an unpredictable path and flattens anything in its way.

While the proposed structure could be built to withstand the impact of a ball of any size, it is not cost-effective to build the structure to withstand the impact of a ball larger than 6 m (19.7 ft). The company insuring the structure requests an analysis to estimate the risk that in any given year the structure will be hit by a ball having a diameter greater than 6 m (19.7 ft).

An engineer is hired to estimate the risk of the proposed structure being hit in any given year by a ball more than 6 m (19.7 ft) in diameter and realizes there are two uncertainties that must be incorporated into the analysis:

(1) the outcome of the roll of the die is unknown; and
(2) the path the ball will follow down the cone is unknown.

Because the probability of the ball taking one path down the cone is the same as the probability of it taking any other path (that is, the probability density is uniform), the probability of the structure being hit is equal to the diameter of the ball divided by the circumferential length at the point where the structure is located. For example, if the diameter of the ball is 6 m (19.7 ft), then the probability the ball will hit the structure is 6/900 = 0.0067, and the probability of a 6 m (19.7 ft) ball being selected is 1/6 = 0.167.

To account for both uncertainties in the analysis, recalling the definition of conditional probability is required. Let $P(A)$ be the probability of event

A occurring, and let $P(B)$ be the probability of event B occurring. Assume events A and B are independent. The probability that event A will occur given that event B has occurred is designated as $P(A|B)$. If $P(AB)$ is the probability that events A and B occur together, then by definition:

$$P(AB) = P(A|B)P(B)$$

Thus, the probability that during any year the proposed structure will be hit by a 6 m (19.7 ft) diameter ball is

$$P(AB) = \left(\frac{20}{3000}\right)\left(\frac{1}{6}\right) = 0.0011$$

The structure will be destroyed if it is hit by a 9, 12, 15, or 18 m (29.5, 39.4, 49.2, or 59.1 ft) diameter ball. Using standard notation, the probability of event C or D occurring is

$$P(C \cup D) = P(C) + P(D) - P(CD)$$

where C is the event a certain number is rolled *and* the structure is hit by the ball and D is the event that a different number is rolled *and* the structure is hit by the ball corresponding to that number. On a single roll of the die two different numbers cannot be obtained; and therefore, $P(CD) = 0$.
    Thus,

$$P(C \cup D) = P(C) + P(D)$$

The probability that the structure will be destroyed in any given year is the sum of the probabilities of it being hit by a 9, 12, 15, or 18 m (29.5, 39.4, 49.2, or 59.1 ft) ball. Let

$$P(hit) = \sum_{k=3}^{6} P_k(hit|D = 3k)$$

or

$$P(hit) = \sum_{k=3}^{6} \left(\frac{10k}{3000}\right)\left(\frac{1}{6}\right) = \frac{10}{3000(6)} \sum_{k=3}^{6} k$$

$$P(hit) = \frac{10}{3000(6)}(3 + 4 + 5 + 6) = 0.01$$

Thus, the 9 m (29.5 ft) ball could be termed the 100-year ball since it is the ball that a diameter or larger that is expected, on the average, at the proposed structure once in 100-years.

Finally, because the probability of destruction of a proposed structure at any given point depends on the circumferential length, for sites with different circumferential lengths there will be 100-year balls of different diameters.

3.4.2 *Implementation for hazard identification*

In discussing implementation, it must be remembered that the purpose of the traditional approach was and is to define regulatory flood hazard zones for insurance purposes; not to provide design guidance. However, it needs to be realized that hazard identification for insurance purposes and design of improvements are inextricably linked; that is, to avoid insurance issues a property (development) in the floodplain needs to be removed from the floodplain by engineering calculation. To accomplish this, modifications of the hazard identification methodology were made and are discussed in the next section.

Let H be a random variable denoting the occurrence of flooding at a specific point on an idealized alluvial fan (Figure 3.1). Let Q be a random variable denoting the peak discharge at the hydrologic apex resulting from a precipitation event in the watershed above the apex. If f_Q is the probability density function of Q at the hydrologic apex, then the probability of a flood

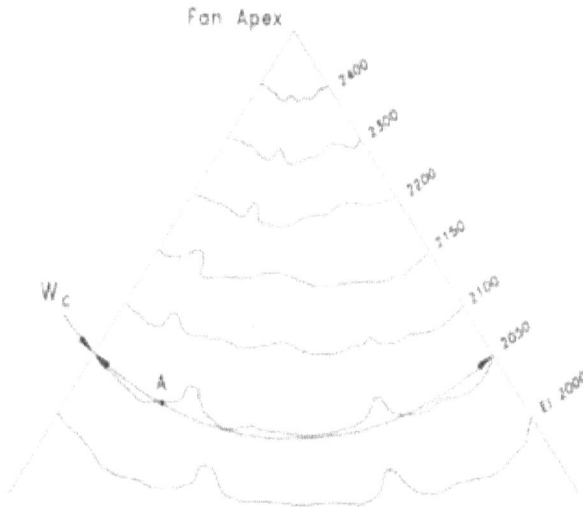

Figure 3.1   Schematic definition of variables to develop a model of flow to a point on an idealized alluvial fan.

with a peak discharge of at least q_0 is

$$P(H = 1) = \int_{q_0}^{\infty} P_{H|Q}(1, q) f_Q dq \tag{3.12}$$

where $P_{H|Q}$ = conditional probability of the point being flooded given that the peak discharge is Q. The 100-year flood at a point is designated as q_{100} and has a 0.01 probability of occurring in any given year where T is the return period in years

$$P(H = 1) = \frac{1}{T} = \frac{1}{100} = \int_{q_{100}}^{\infty} P_{H|Q}(1, q) f_Q(q) dq \tag{3.13}$$

At this point it is appropriate to observe that Eq. (3.12) can be solved for any return period although 100-years is the typical regulatory requirement. If the traditional approach is valid for events with a return period of 100-years, then it should be valid for events with greater return periods; however, careful consideration must be given for the validity of events with smaller return periods, for example 25-years. Recall, one of the critical assumptions is that the flow has sufficient energy to form its own channel and smaller events may not have the energy required to accomplish this.

The conditional probability $P_{H|Q}(1, q)$ is taken as the ratio of the width of the channel formed by the flow and the width of the alluvial fan contour on which the point of interest lies (Dawdy, 1979) or

$$P_{H|Q} = \frac{w(q)}{W_c} \tag{3.14}$$

where $w(q)$ = the width of the channel at flow rate q and W_c = alluvial fan contour width. If $f_Q(q)$ in Eq. (3.13) is an LP3 type distribution then it can be shown that when terms in the integral are combined the following equation results

$$P(H = 1) = \frac{1}{T} = \frac{1}{100} = \frac{A\beta C}{W_c} \int_{z_{100}}^{\infty} f_z'(z) dz \tag{3.15}$$

where $f_z'(z)$ is a transformed LP3 distribution with a mean value $\overline{Z}$ (Eq. 3.8), a standard deviation S_z (Eq. 3.8), skew G_z (Eq. 3.10), and β is a coefficient equal to 9.41 in the single channel region and 35.8 in the multiple channel region, and C is a transformation coefficient (Eq. 3.8). It is also standard practice to introduce an avulsion coefficient (A) to take into account the possibility of the flow abandoning its current channel and forming or capturing a "new" channel. As noted previously, A varies between 1 and 2 with a generally conservative value of 1.5 generally being used.

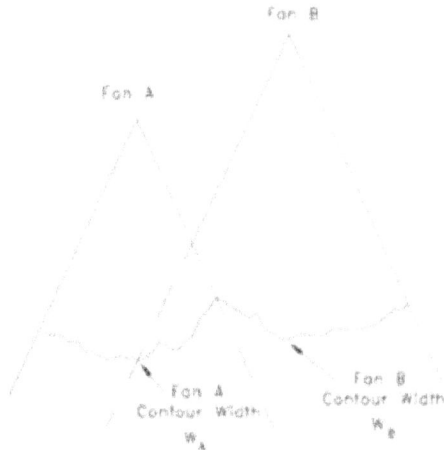

Figure 3.2   Schematic definition of variables on idealized coalescent alluvial fans.

In many cases of interest, a point on an alluvial fan may be subject to flood waters from more than one hydrologic fan apex; that is, the point is located on coalescent alluvial fans (Figure 3.2). In this situation, as noted above, that flooding at the point of interest by water from one fan apex is independent of the event of flooding at this point from the other fan apex; therefore, it is the union of these events that defines the probability of flood at the point of interest or

$$P(A \cup B) = P(A) + P(B) - P(A \cap B)$$

In practice, events A and B are assumed to be independent, which is technically reasonable; and if this is the case, then

$$P(A \cap B) = P(A)P(B)$$

In the most cases, the event with a 100-year return period, or the event with an exceedance probability of 0.01, is the event of interest; that is

$$P(A \cup B) = 0.01$$

This implies that $P(A) \leq 0.01$ and $P(B) \leq 0.01$; and therefore, $P(A \cup B)$ has a maximum value when $P(A) = P(B) = 0.005$ or

$$P(A \cap B) = P(A)P(B) = (0.005)(0.005) = 2.5 \ x \ 10^{-5}$$

and this term is neglected. Therefore, in the case of coalescent alluvial fans the model for estimating the flow of a specified return period to a point on the fan surface is

$$\frac{1}{T} = \frac{\beta_A A_A C_A}{W_A} P_A + \frac{\beta_B A_B C_B}{W_B} P_B \tag{3.16}$$

where the subscripts identify the fan surface (Figure 3.2) to which the parameter applies and all variables are as previously defined.

3.5   An Approach to Hazard Mitigation

While the foregoing section addresses how flood hazards on alluvial fans may be estimated from a regulatory viewpoint, it does not suggest how peak flood flow to a structure or development; that is an object of finite width on an alluvial fan, should be estimated. From the viewpoint of civil engineering this is an essential issue. There are two other missing elements to the approach outlined in the previous section. First, as previously mentioned, on-fan precipitation and infiltration are not taken into account. Prudent development on an alluvial fan suggest that both of these must be taken into account for the design of an adequate drainage system to protect a structure on the fan. Second, sediment transport is not taken into account. Both of these considerations add considerable complexity to the problem. The first element involves understanding both the surficial and sub-surface geology, which is addressed in Chapter 2. The second element is difficult; that is, the sediment load being transported by the flow must be estimated and then when this flow impacts the structure its pattern of deposition must be estimated. The pattern and depth of deposition can be theoretically be estimated by the method developed by French *et al.* [2001]; however, this approach has not been verified either experimentally or with field measurements.

To estimate the peak flow to a structure of finite width on an alluvial fan, consider a structure that has length W_s projected onto an alluvial fan contour. Then, Eq. (3.14) becomes

$$P_{H|Q} = \frac{w(q) + W_s}{W_c} \tag{3.17}$$

With this formulation, Eq. (3.15) becomes

$$P(H = 1) = \frac{1}{T} = \frac{1}{100} = \frac{A\beta C}{W_c} \int_{100}^{\infty} f_z'(z)dz + \frac{W_s}{W_c} \int_{100}^{\infty} f_z(z)dz \tag{3.18}$$

where $f_z'(z)$ is the transformed LP3 distribution and $f_z(z)$ is the untransformed LP3 distribution. This development can easily be extended to coalescent where the structure is in the coalescent area or following the form of Eq. (3.16)

$$\frac{1}{T} = \frac{\beta_A A_A C_A}{W_{cA}} P_{1A} + \frac{W_{sA}}{W_{cA}} P_{2A} + \frac{\beta_B A_B C_B}{W_{cB}} P_{1B} + \frac{W_{sB}}{W_{cB}} P_{2B} \qquad (3.19)$$

where P_{1A} and P_{1B} are the transformed exceedance probabilities associated with fans A and B, respectively; and P_{2A} and P_{2B} are the untransformed exceedance probabilities associated with fans A and B, respectively.

3.6   Conclusion

While the traditional methods of flood hazard and mitigation on alluvial fans have both their theoretical and practical limitations, there are several observations that must be made. First, those responsible for identifying flood hazard and mitigating it on alluvial fans recognized that flooding on alluvial fans presented problems that could not be solved using traditional hydrologic and hydraulic engineering methods that are effective in riverine environments [Edwards and Thielman, 1984]. Therefore, in response to an identified unique problem a first generation approach was developed and in many ways it was prescient in its insight to the problem; see for example, Cazanacli *et al.* [2002] and French and Miller [2003]. Many of the assumptions put forward by Dawdy [1979] were and remain open issues; however, a critical engineering issue was identified and a solution found. Second, in general, the traditional methodology results in conservative answers and regulators whether at the federal, state, or local levels favor conservative answers since in many ways they are the ultimate insurers of public safety and property. This statement is supported by published research studies; for example, Flippin and French [1994] and Miller and French [1995 and 1996]. Third, other stochastic models have been proposed, Heggen and Anderson [1994] and there have been efforts to apply existing deterministic models to the problem, Mustaq and Mays [1991] with varying degrees of success.

This chapter has been a brief summary of the beginning of dealing with the "traditional" approach to the identification of flood hazard and mitigation on alluvial fans and is by no means complete. For example, space considerations did not allow for the presentation of examples (the reader is referred to French *et al.* [1996] for a report that is replete with examples) and also a full discussion of the impact of sediment transport on flood hazard and mitigation on alluvial fans; however, sediment transport issues are fully covered in Garcia [2008]. At this point the readers' attention is called to the paper by Zhao and Mays [1996], which quantifies the risk and uncertainty inherent in using the traditional approach. In the next chapter, current approaches are presented and discussed and all approaches are illustrated and further discussed in the chapter presenting case studies.

References

Bull, W.B. (1964). "History and causes of channel trenching in western Fresno County, California." *American Journal of Science*, 262, 249–258.

Cazanacli, D., Paola, C. and Parker, G. (2002). "Experimental steep braided flow; Application to flooding risk on fans." *Journal of Hydraulic Engineering*, 128(3), 322–330.

Dawdy, D.R. (1979). "Flood frequency estimates on alluvial fans." *Journal of the Hydraulics Division*, 105(HY11), 1407–1412.

Edwards, K.L. and Thielman, J. (1984). "Alluvial fans: a novel flood challenge." *Civil Engineering*, 54(11), 66–68.

FEMA (1983). "Appendix G: alluvial fan studies." *Guidelines and Specifications for Study Contractors*. FEMA 37, Federal Emergency Management Agency, Washington, DC.

FEMA (1990). "FAN: an alluvial fan flooding computer program." Federal Emergency Management Agency, Washington, DC.

Flippin, S.J. and French, R.H. (1994). "Comparison of results from an alluvial fan design methodology with historic data." *Journal of Irrigation and Drainage Engineering*, 120(1), 195–210.

French, R.H. (1992). "Preferred directions of flow on alluvial fans." *Journal of Hydraulic Engineering*, 118(7), 1002–1013.

French, R.H. and Miller, J.J. (2006). "El Nino — La Nina implications on flood hazard mitigation." Proceedings of the World Water and Environmental Resources Congress, ASCE/EWRI, Omaha, Nebraska.

French, R.H. and Miller, J. (2003). Discussion of "Experimental steep braided flow; Application to flooding risk on fans." by D. Cazanacli, C. Paola, and G. Parker, *Journal of Hydraulic Engineering*, 128(3), 2002, 129(11), 920–922.

French, R.H. and Miller, J.J. (2002). Analysis of a Design Level Precipitation Event in Area 3 of the Nevada Test Site, Report No. 45198, Division of Hydrologic Sciences, Desert Research Institute, Las Vegas, NV.

French, R.H., Miller, J.J. and Curtis, S. (2001). "Estimating the depth of deposition (Erosion) at slope transitions on alluvial fans." *Journal of Hydraulic Engineering*, 127(9), 780–782.

French, R.H., McKay, W.A. and Fordham, J.W. (1996). *Chapter 3: Identification and Mitigation of Flood Hazard on Alluvial Fans*, unpublished report for U.S. Department of Energy, available for copying at Desert Research Institute Library (call no. 29-UN1DC/9: I3), Las Vegas, NV.

French, R.H., Fuller, J.E. and Waters, S. (1993). "Alluvial fan: Proposed new process-oriented definitions for arid southwest." *Journal of Water Resource Planning and Management*, 119(5), 588–598.

Garcia, M. (2008). *Sedimentation Engineering: Processes, Measurements Modeling, and Practice*, Garcia, M.H. (ed.), ASCE Manual of Practice 110, Reston, VA.

Garcia, M., MacArthur, R., French, R. and Miller, J. (2008). "Sedimentation hazards." *Sedimentation Engineering: Processes, Measurements, Modeling, and Practice*, Garcia, M.H. (ed.), ASCE Manual of Practice 110, Reston, VA., 885–922.

HEC (1990). "HEC-1, flood hydrograph package, user's manual." U.S. Army Corps of Engineers, Hydrologic Engineering Center, Davis, CA.

HEC (2000). Hydrologic modeling system, HEC-HMS." U.S. Army Corps of Engineers, Hydrologic Engineering Center, Davis, CA.

Heggen, R.J. and Anderson, R.W. (1994). "Probabilistic modeling of floodplain avulsions." *Colorado Association of State Floodplain Managers*, Telluride, CO.

Hooke, R. LeB. (1965). "Alluvial fans." Ph.D. Dissertation, California Institute of Technology, Pasadena, CA.

Hooke, R. LeB. (1967). "Processes on arid-region alluvial fans." *American Journal of Science*, 266, 609–629.

Leopold, L.B. (1951). "Rainfall frequency: an aspect of climatic variation." *Transactions of the American Geophysical Union*, 32(3), 347–357.

Magura, L.M. and Wood, D.E. (1980). "Flood hazard identification and flood plain management on alluvial fans." *Water Resources Bulletin*, 16(1), 56–62.

Miller, J.J. and French R.H. (1996). "Mitigation of flood hazards on alluvial fans" *North American Water and Environment Congress96*, ASCE, Anaheim, CA.

Miller, J.J. and French, R.H. (1995). "Evaluating alluvial fan flood hazards along rail corridors to the Department of Energy's Proposed High-Level Radioactive Waste Disposal Site at Yucca Mountain." *Geological Society of America Annual Meeting Proceedings*, New Orleans, LA.

Mushtaq, H. and Mays, L.W. (1991). "Hydraulic modeling of alluvial fans using DAMBRK (NWS computer model)." HPR-PL-1(37), Arizona Department of Highways, Phoenix, AZ.

Reich, B.M. and Renard, K.G. (1990). "Graphic advances aid flood engineers." *Hydraulics/Hydrology of Arid Lands, Proceedings of the International Symposium*, French, R.H. (ed.), ASCE, Reston, VA, 737–742.

Reich, B.M., Renard, K.G. and Lopez, F.A. (1990). "Alignment of large flood-peaks on arid watersheds." In: *Hydraulics/Hydrology of Arid Lands, Proceedings of the International Symposium*, French, R.H. (ed.), ASCE, Reston, VA, 477–482.

Schick, A.P. (1974). "Alluvial fans and desert roads — a problem in applied geomorphology." *Abhandlungen der Akademie der Wissenshaften* in Gottingen *Mathematisch-Physikalishe Klasse*, 29, 418–425.

Schumm, S.A., Baker, V.R., Bowker, M.F., Dixon, J.R., Dunne, T., Hamilton, D., Hjalmarson, H.W. and Merritts, D. (1996). *Alluvial Fan Flooding*, Committee on Alluvial Fan Flooding, Water Science and Technology Board, Commission on Geosciences, Environment, and Resources, National Research Council, National Academy Press, Washington, D.C.

Thomas, B.E., Hjalmarson, H.W. and Waltemeyer, S.D. (1994). "Methods for estimating magnitude and frequency of floods in the Southwestern United States." Open File Report 93-419, U.S. Geological Survey, Washington, D.C.

WRC (1981). "Guidelines for determining flood flow frequency." *Bulletin 17b*, U.S. Water Resources Council, Washington, D.C.

Zhao, B. and Mays, L. (1996). "Uncertainty and risk analyses for FEMA alluvial-fan method." *Journal of Hydraulic Engineering*, 122(6), 325–332.

Chapter 4

New Approaches for Alluvial Fan Flood Hazard

Jimmy S. O'Brien

FLO-2D Software, Inc.
P.O. Box 66, 102 County Road 2315, Nutrioso, Arizona 85932
jim@flo-2d.com

Reinaldo Garcia

Applied Research Center, Florida International University
10555 West Flagler Street EC 2100, Miami, Florida 33174
reinaldo@flo-2d.com

Beginning with a review of FEMA's three phase approach, this chapter discusses numerical methods for predicting alluvial fan flooding with the focus on deterministic methods for unconfined flow. The limitations of the FEMA FAN model are outlined as background for discussing the important criteria for delineating fan flood hazards. Alluvial fan flood modeling requires an understanding of the potential sediment loading, flow path uncertainty, and flooding details such as rainfall, urban impacts, and hydraulic controls. An approach used worldwide for delineating flood hazards is recommended and some important considerations for alluvial fan flood mitigation are presented.

4.1 Predicting Alluvial Fan Flooding — Background

Alluvial fan flooding is distinguished by unconfined flow, steep slope and high sediment transport. The steep slope and inexhaustible sediment supply create flow path uncertainty. The nature of alluvial fan evolution requires that the flooding distributes the sediment more or less uniformly over the fan through geologic time, but flood hazard delineation, regulation and mitigation occur in engineering time scales. Engineers face the dilemma of predicting fan flood hazards for both existing conditions and potential future topographic conditions. This dichotomy has led to two approaches, stochastic or risk-based approximate methods and deterministic models using existing conditions. The focus of this chapter is advanced numerical modeling methods for alluvial fans. FEMA's evolving approach to alluvial fan mapping leads this discussion because of its widespread application for the purpose of assigning flood insurance rates.

The stochastic analysis of alluvial fan modeling began with Dawdy's [1979] publication on flood frequency estimates on alluvial fans that eventually became the basis of the FEMA FAN modeling method. The inherent assumption with this method was that each flood event forms a random single channel and the flow stays in that channel throughout the flood even if the channel avulses. The random flood channel is then assumed to be uniformly distributed across any contour. The prescribed method estimated the probability of a fan point along a given contour and computed the width of a contour using the channel width. Mifflin [1988] expanded the 'Dawdy Method' to computed design depths and velocities on alluvial fans using the size of the area at risk. He further demonstrated that the method could be used on entrenched channels at the fan apex [Mifflin, 1990]. Using available data, French [1991] determined that the probability of a point on the fan being flooded was better represented by a normal distribution than a uniform probability.

Stochastic methods for delineating alluvial fan flood hazard zones used by FEMA rely on the concept of idealized fan morphology where through geologic time channels migrate over the surface to create a symmetric fan shape. The inherent assumption is that the channel location is random. The difficulties with this stochastic approach are:

(1) Most FEMA Flood Insurance Studies (FIS) are undertaken because the fan is already urbanized or development is planned. Urban drainage facilities eliminate the random location of channels on the fan.
(2) Even at the fan apex, most alluvial fan channels evolve in response to the 2- to 5-year return period flood event. The design flood event is typically the 100-year flood and most of the flooding for this event at the apex is unconfined (not channel conveyance). The pre-flood channel geometry is often destroyed.
(3) The stochastic method does not lend itself to accurate flood hazard mapping. Nor can the stochastic method accurately estimate spatially

variable inundation and the resultant flood probability if urban conditions are not taken into account. Advanced numerical methods for addressing alluvial fan flood risk will be discussed in the Chapter 5. A discussion of available flooding routing models follows.

Estimating unconfined flood hydraulics on alluvial fans is a relatively recent science. Most of the early attempts to predict fan flood hydraulics involved developing the shear stress equations for mud and debris flows. Yano and Daido [1965] applied the Bingham rheological model for clay slurries in open channels. Dispersive stress (particle contact) theory was advanced by Bagnold in 1954 and was later applied by a number of researchers to predict open channel hydraulics for hyperconcentrated sediment flows [Lowe, 1976; Takahashi, 1978, 1980; Savage, 1979; and Savage and McKeown, 1983]. The development of mathematical models for routing non-Newtonian flows in open channels were initiated by DeLeon and Jeppson [1982], Schamber and MacArthur [1985] and Takahashi and Tsujimoto [1985]. O'Brien [1986] designed a one-dimensional mudflow model for watershed channels that also utilized the Bingham model. In 1985, Schamber and MacArthur presented a two-dimensional finite element model for application to simplified overland topography [USCOE, 1988]. Later Takahashi and Nakagawa [1989] modified their debris flow model to include turbulence.

Numerical models can now simulate alluvial fan flood physical processes with increasing accuracy and resolution. Complex phenomena such as hyperconcentrated sediment flows (mud and debris flows), distributary channels and structural mitigation form the basis of many engineering fan flood studies. Virtually all alluvial fan flood hazard delineation studies require two-dimensional flood routing models applying the full dynamic wave momentum equation. The key component of these flood routing models is the volume conservation of the watershed hydrograph.

While technical publications with research on advanced numerical flood routing methods are prolific, these have marginal practical use for the engineer and floodplain manager. There are only a limited number of commercially available two-dimensional flood routing models. These include the MIKE FLOOD from Danish Hydraulic Institute (DHI); FLO-2D® from FLO-2D Software, Inc.; TUFLOW from BMT WBM Pty. Ltd.; SOBEK from Delft Hydraulics Software; InfoWorks™RS from Wallingford Software; and two older models, the USCOE RMA2 and the USDOT FESWMS-2DH. Of these models only the MIKE FLOOD, FLO-2D, RMA2 and

FESWMS-2DH are on FEMA's list of approved hydraulic models for two-dimensional FIS studies and only MIKE FLOOD (2005) and FLO-2D (2007) are identified by FEMA as approved hydraulic models for alluvial fan studies. Both models offer regular update modeling enhancements. Since most fan flood analyses are for FEMA FIS or for possible removal from the mapped floodplain, it is prudent to begin a discussion of fan hazard mapping with FEMA's current approach.

4.2   FEMA's Three Phase Approach to Alluvial Fan Flood Mapping

Until 1990, virtually all alluvial fan flood hazard studies in the United States that did not involve mud and debris flows utilized either the Corps of Engineers HEC-2 steady discharge, river backwater model or the FEMA FAN stochastic model [FEMA, 2000]. The HEC-2 model was used in most cases simply because engineers did not have any other modeling tools. It is not necessary to elaborate on the obvious difficulties with using a steady discharge model based on a 1-D solution to the energy equation to predict water surface elevations on a steep sloped and variable topography alluvial fan. The FEMA FAN model to predict alluvial fan flooding took the opposite approach and arose from the concept that the flow path uncertainty is 'so great' that it "...cannot be set aside in the realistic assessment of flood risk or in the reliable mitigation of the hazard." This concept presumed that the area of inundation was unpredictable due to flow path avulsion. As a result, FEMA's 'Guidelines and Specifications for Flood Hazard Mapping Partners, Appendix G Guidance for Alluvial Fans Flooding Analyses and Mapping' [FEMA, 2002] stipulates that the unpredictable or 'ultra hazardous' fan flooding condition renders the potential flood mitigation based on conventional floodplain fill construction techniques as unreliable.

To address fan flood hazard mapping for insurance purposes, FEMA embraced a simple stochastic method that essentially delineated the entire alluvial fan as a flood hazard. The FAN model, as it is referred to, has been used extensively for FIS on alluvial fans. It has also been used in the past, in the absence of other applicable predictive methods, to estimate flood hydraulics (depth and velocity) for both local agency regulation and mitigation. The FAN method is discussed in the FEMA Guidelines [FEMA, 2002]. The FAN program results include estimates of the mean depth and velocities on radial segments of the fan (delineated as insurance rate zones),

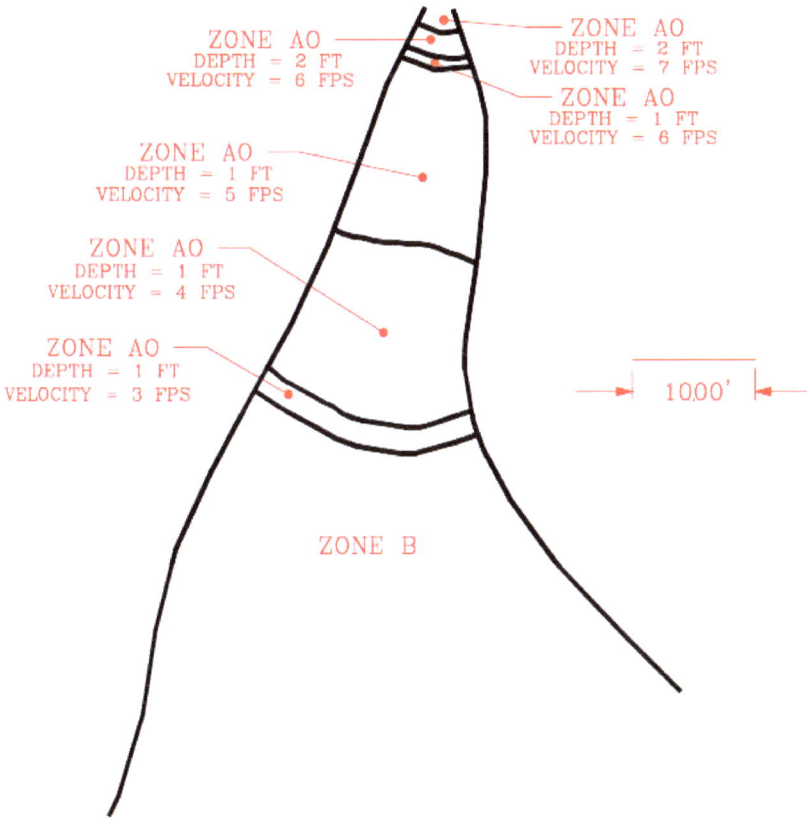

Figure 4.1   Example: FEMA FAN Model flood insurance rate zones [from FEMA, 2002].

limited by user imposed topographic width boundaries (Figure 4.1). The hydrologic data requirements are at least three peak discharges and the corresponding frequency of occurrence (in terms of the return period events; *e.g.*, 10-year, 50-year and 100-year floods).

The inherent assumptions in the FAN model provide some insight into floodplain management perceptions of fan flood hydraulics for later discussion on fan flood risk analyses. These assumptions are:

(1) Flow hydraulics are based on 'normal' depth inferring that the average flow condition is both steady and uniform;
(2) Normal depth flow is critical, providing an additional simplifying equation to predict the flow width;

(3) Flow is confined to a multiple channel or distributary channel of a prescribed width as function of the discharge; and,

(4) The probability of the channel location on the fan is uniformly distributed along the topographic contour. This infers uniformity of surface topography and roughness along a given contour between limits of the user defined fan surface.

The limitations of the FEMA FAN approach are apparent and include:

(1) Failure to consider floodwave attenuation;

(2) Variable surface conditions including topography and roughness result in unsteady and non-uniform flow;

(3) Mobile bed flow is inherently subcritical;

(4) Neglect of local fan rainfall and infiltration;

(5) Failure to consider urban impacts;

(6) Failure to assess migration measures and other hydraulic controls such as bridges and culverts; and,

(7) The important consideration from a mapping perspective is that flooding on alluvial fans, as for most flood studies, is controlled by the hydrograph volume (at the fan apex) not the peak discharge. On some long alluvial fans, floodwave attenuation can be so significant that the entire flood can be dissipated before the fan terminus is reached.

To counter some of these limitations and difficulties, FEMA developed a three phase approach that included an investigation of fan geomorphology, mapping of active and inactive areas, and the 100-year flood hazard delineation in their publication of the 2002 Guidelines. The three-phase approach is an attempt to combine the geomorphic concept of fan evolution with deterministic fan modeling methods and is briefly outlined so that reference can be made to these phases later.

4.2.1 *Identification of fan geomorphology*

As a first step FEMA indicates that it is critical to recognize and characterize the landform as an alluvial fan. This is necessary to apply the appropriate methods for flood hazard delineation. Fan composition, morphology, and location are discussed as FEMA guidelines for fan classification. Composition refers to the fan alluvium or sediment delivered by flooding or debris flows from the upper watershed. Morphology considers on the alluvial fan drainage pattern associated with avulsing distributary flows to evolve the conical fan shape. Location as an index is defined by the recognition of the fan hydrologic apex which is marked by a distinct break in the topographic slope. The fan geomorphic apex is distinguished from the hydrologic apex which is defined by where channel bifurcation initiates and flow may become unconfined.

4.2.2   *Active versus inactive fan areas*

Delineating active versus inactive fan areas is an attempt to isolate the portions of alluvial fan that may appropriate for development. This delineation is based on the recognition of fan processes including deposition, erosion and unstable channels contributing to the fan morphology. While inactive portions of the fan may be flooded in future events, defining the active portions through soil composition, vegetation, and exposed geology will help to identify areas that are clearly susceptible to current flooding. Basically if flooding and sediment deposition have occurred on a portion of an alluvial fan, it is considered to be active. This analysis of the spatial extent of inundation is a qualitative analysis that should be supported by hydraulic computations. Delineated inactive fan areas may be associated with channel or wash incision.

4.2.3   *100-year flood hazard modeling and mapping*

The delineation of alluvial fan flood hazard (Special Flood Hazard Areas; SFHA) for mapping purposes is based on the 100-year flood. FEMA suggests that the variability of land forms and flood processes requires a flexible approach to alluvial fan flood mapping. Several predictive methods are mentioned in the FEMA Guidelines [2002] including the FEMA FAN model previously discussed. For the engineer applying a deterministic model, the FEMA guidelines suggest that two-dimensional models are useful for simulating alluvial fan flooding where flows have high sediment loading, split flows, mud and debris flows, or complex urban flooding. Taking this FEMA initiative, two-dimensional alluvial fan flood modeling is reviewed.

4.3   Alluvial Fan Flood Modeling

Most numerical flood routing models use volume conservation and the momentum equations to distribute a flood over an unconfined surface. Over the last decade, these numerical models have become increasingly faster, stable, and more reliable. To highlight one of these models used extensively to predict alluvial fan flooding by both agencies and consultants, the discussion will focus on the FLO-2D model. FLO-2D has a number of unique features for simulating alluvial fan flooding including the ability to simulate mud and debris flows, the option to model distributary channels, street flow and flow obstructions, and the opportunity to model rainfall and infiltration during the flood event. The options for simulating fan flooding with FLO-2D include:

(1) Main channels with surveyed cross section data;
(2) Levees (berms) and levee breach failure;
(3) Detention basins;
(4) Culverts and bridges;
(5) Street flow;
(6) Buildings and other obstructions;
(7) Sediment transport; and,
(8) Diversions.

4.3.1 *Developing an alluvial fan flood model*

The formation of most numerical models require the same steps regardless if the model is applied to alluvial fans or floodplains. For either finite element or finite difference numerical models, the two critical ingredients are topography and hydrology. Gradually most of the western United States and many of the urban areas around the world are being flown for digital terrain data as either Light Detection and Ranging (LIDAR) or X-Band Radar Data. This topography data is provided as a set of random elevation points with associated coordinate geometry as a Digital Terrain Elevation Model (DEM or DTM). Typically the DTM data is imported into a pre-processor program with aerial imagery in the background and a grid system is overlaid. The grid element or cell is then assigned a representative elevation based on interpolation and filtering of the DTM points. The interpolation may be distance weighted from the cell center. Filtering the DTM data enables a more representative floodplain surface to be assigned where there are buildings, structures or dense vegetation.

The required hydrology can take the form of an inflow hydrograph at the fan apex or a rainfall-runoff simulation. In FLO-2D, any number of inflow hydrographs can be assigned to model coalescing alluvial fans. Rainfall can be spatially and temporally variable over a combined model of the upstream

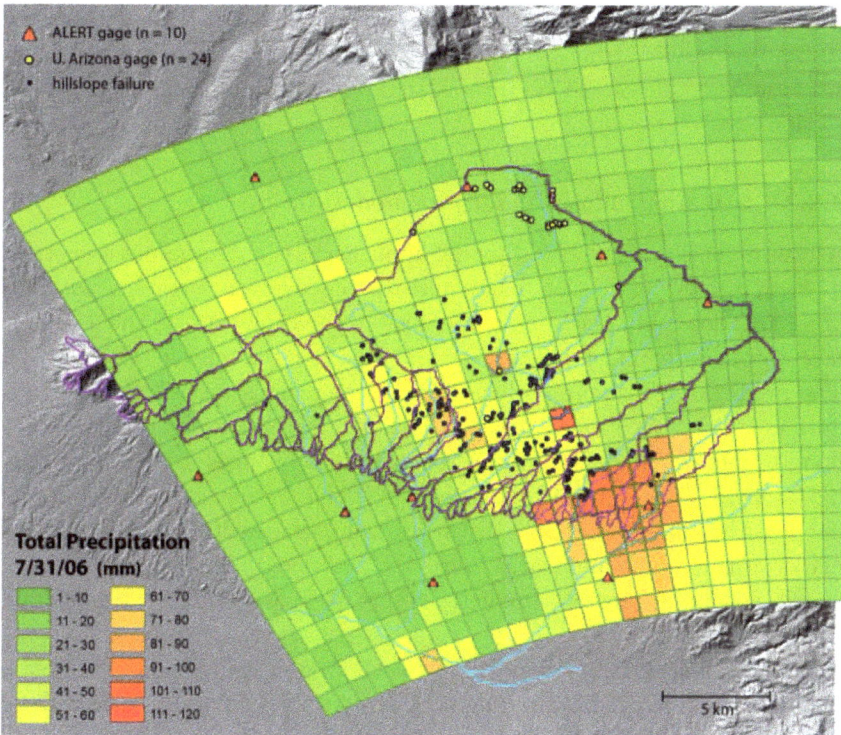

Figure 4.2 NEXRAD cell total precipitation July 31, 2006, Catalina Mountain watersheds, Southeastern Arizona (USGS Email August 7, 2007).

watershed and alluvial fan. Doppler (NEXRAD) radar data or any gridded data base can be used to simulate actual storm events with spatial variability (Figure 4.2). Rainfall runoff infiltration loss can be computed with either the Green-Ampt or SCS infiltration models. When the alluvial fan surface area is on the same order of magnitude as the watershed, then the rainfall on the fan can be an important contribution to the overall flood volume. The rainfall can be added to the floodwave as it progresses over the alluvial fan or it can wet the surface and fill the infiltration storage prior to arrival of the frontal wave.

A basic alluvial fan rigid bed FLO-2D flood model consists of a grid system with assigned grid element elevations and hydrology in the form of an inflow hydrograph or rainfall. With the GDS integrated preprocessor, this interactive data input can take just a few minutes to set up. The flood simulation resolution can be enhanced by adding details to the model. First the main channel should be added, then streets, levees and hydraulic structures. Finally, mobile bed with erosion and deposition and mudflows can be investigated which will be a focus of a later section.

4.3.2 *2-D unsteady alluvial fan model limitations*

A two-dimensional numerical model application is a deterministic method of flood hazard prediction based on existing or design conditions. This approach does not encompass the long term evolution of the fan surface through sediment deposition and flow path avulsion (Chapter 1). The prediction of the area of inundation for various return period floods does not directly consider either the change in topography due to recent flood events or the uncertainty associated with changes in flow path during a flood event. A series of frequent flood events or a large design flood can result in a severely disturbed fan surface with an incised channel or wash or an apex area buried in sediment. Subsequent flooding would not be accurately predicted using previous fan topography and roughness.

For flood hazard mapping, it is not the hydraulic engineer's or floodplain manager's task to predict the long term geomorphic evolution of the alluvial fan. This would require simulating potentially hundreds or thousands of flood events including modeling sediment deposition and scour. This is an impossible task because it is the sequence of the flood events and associated sediment loading that will alter the fan topography near the apex from one event to another. To give the modeler some confidence to tackle fan flooding, there are several issues that simplify the task:

(1) Fan apex channels will have a capacity on the order of the 2- to 5-year return period flood assuming that the channel is not incised.
(2) Infrequent floods such as the 100-year flood are so large that the apex channel capacity is not critical to the prediction of the area of inundation.
(3) For large floods in the absence of severely disturbed surface conditions, flood stage water surface elevations are more sensitive to the overall flow roughness than the local aggradation/degradation of the bed [National Research Council, 1983].

To overcome the inherent uncertainty of the potential flow path using a deterministic model, there are several tools at the modeler's disposal to estimate the maximum potential area of inundation. These include:

(1) Determining flow path sensitivity to topography;
(2) Forcing different flow path directions at the fan apex; and,
(3) Sediment transport analyses (mobile bed simulation).

Flow path uncertainty will be addressed in more detail in the following chapter, but the issue of sediment transport and mudflows will be discussed to demonstrate the complexities and limitations associated with sediment modeling.

4.3.3   *Alluvial fan sediment issues*

Depending on geology and soil conditions in the upper watershed, sediment loading at the fan apex can encompass the continuum from clear water flows to landslides. There are two primary issues related to sediment deposition on the alluvial fan from an engineering perspective:

(1) Accurate estimates of sediment yield from the upper watershed; and,
(2) Prediction of sediment transport capacity on the fan surface.

Similar to flood water volume, the area of inundation and impacts on the alluvial fan from sediment deposition are a function of the sediment yield estimate from the basin. There are three primary sources of sediment:

(1) Overland sediment yield;
(2) Hillslope sloughing and landslide failure; and,
(3) Channel bed and bank erosion.

Within the context of the geology and watershed condition, each source of sediment can be evaluated by the engineer to estimate the potential sediment loading to the fan. Fires, land use practices, and other disturbances can exacerbate the sediment yield. Estimating the sediment yield is an inexact science, but its relationship to the water volume in the design flood event can improve the accuracy of the flood hazard.

For flooding purposes, the concept of fluid behavior is limited to a maximum concentration by volume on the order of 55 to 60 percent (Table 4.1). For higher sediment concentrations, the earth movement should be classified as a landslide and an analysis of geotechnical slope stability failure would be required. The ability of a watershed to generate a mudflow is primarily dependent on the type and concentration of the fine sediment

Table 4.1   Hyperconcentrated sediment flow behavior.[1]

	Sediment concentration		
	by Volume	by Weight	Flow characteristics
Landslide	0.65–0.80	0.83–0.91	Will not flow; failure by block sliding.
	0.55–0.65	0.76–0.83	Block sliding failure with internal deformation during the slide; slow creep prior to failure.
Mudflow	0.48–0.55	0.72–0.76	Flow evident; slow creep sustained mudflow; plastic deformation under its own weight; cohesive; will not spread on level surface.
	0.45–0.48	0.69–0.72	Flow spreading on level surface; cohesive flow; some mixing.
Mud flood	0.40–0.45	0.65–0.69	Flow mixes easily; shows fluid properties in deformation; spreads on horizontal surface but maintains an inclined fluid surface; large particle (boulder) setting; waves appear but dissipate rapidly.
	0.35–0.40	0.59–0.65	Marked settling of gravels and cobbles; spreading nearly complete on horizontal surface; liquid surface with two fluid phases appears; waves travel on surface.
	0.30–0.35	0.54–0.59	Separation of water on surface; waves travel easily; most sand and gravel has settled out and moves as bedload.
	0.20–0.30	0.41–0.54	Distinct wave action; fluid surface; all particles resting on bed in quiescent fluid condition.
Water flood	<0.20	<0.41	Water flow with conventional suspended load and bedload.

[1]From O'Brien [1986].

(silts and clays) in the fluid matrix. Typically desert alluvial fans will have a minor silt and clay size fraction whereas as mudflow deposit will most likely have clay content above 10 percent.

Desert alluvial fan floods without a significant clay fraction will have sediment concentrations in the range from 5 to 20 percent concentration by volume. For fan flooding, the hyperconcentrated sediment particle distribution in the flow profile is more uniform than conventional sediment transport involving bedload and suspended load processes. The application of conventional riverine sediment transport equations is clearly inappropriate, but unfortunately the development of sediment transport capacity equations for steep slopes with high sediment concentrations and inhibited particle fall velocities are limited to only a couple of extended conventional river equations.

There are nine river sediment transport equations in the FLO-2D model. Although three equations have been identified for steeps slopes, MPM-Smart [Smart, 1984], Zeller-Fullterton [1983] and MPM-Woo [see Mussetter *et al.*, 1994; and Woo *et al.*, 1988], validation of these equations for alluvial fan conditions has not been undertaken. Two other equations, Ackers and White [1973] and Englund and Hansen [1967] will also predict high sediment transport rates. These five equations may generate sediment loads as high as 10 percent concentration by volume (or higher in the case of MPM-Woo), but this is still less than the threshold of 20 percent concentration by volume for hyperconcentrated sediment flows (Table 4.1). A brief description of the nine equations follows:

Engelund-Hansen Method. Bagnold's [1954] stream power concept was applied with the similarity principle to derive a sediment transport function. In accordance with the similarity principle, the method should be applied only to flow over dune bed forms, but Engelund and Hansen (1967) determined that it could be effectively used in both dune bed forms and upper regime sediment transport (applicable for alluvial fans) for particle sizes greater than 0.15 mm.

Karim and Kennedy. This method provides the best replication of field data for rivers compared to all other total load equations and has been developed and applied for size fractions [Karim, 1988].

Laursen's Transport Function. The Laursen [1958] formula agreed well with field data from small rivers. For larger rivers the correlation between measured data and predicted sediment transport was poor [Graf, 1971]. This set of equations involved a functional relationship between the flow hydraulics and sediment discharge.

MPM-Smart Modification. Smart's [1984] investigation of the Meyer-Peter Mueller equation bedload equation results in modifications to predict sediment transport for slopes up to 20 percent and for mean grain size greater than 0.4 mm.

MPM-Woo Relationship. For computing the bed material load in steep sloped, sand bed channels such as arroyos, washes and alluvial fans, Mussetter *et al.* [1994] linked Woo's relationship for computing the suspended sediment concentration with the Meyer-Peter-Mueller bedload equation. Woo *et al.* [1988] developed an equation to account for the variation in fluid properties associated with high sediment concentration. Mussetter *et al.* [1994]

derived a multiple regression relationship to compute the bed material load for a range of hydraulic and bed conditions typical of fans in the Southwest. This equation provides a method for estimating high bed material load in steep, sand bed channels that are beyond the hydraulic conditions for which the other sediment transport equations are applicable.

Toffeleti's Approach. Toffaleti [1969] developed a procedure to calculate the total sediment load by estimating the unmeasured load. Following the Einstein approach, the bed material load is given by the sum of the bed-load discharge and the suspended load in three separate zones. Toffaleti computed the bedload concentration from his empirical equation for the lower-zone suspended load discharge and then computed the bedload.

Yang's Method. Yang [1973] determined that the total sediment concentration was a function of the potential energy dissipation per unit weight of water (stream power) and the stream power was expressed as a function of velocity and slope. The majority of the data used to develop Yang's method was limited to medium to coarse sands and flow depths less than 3 ft (1 m) [Julien, 1995]. Yang's equations in the FLO-2D model can be applied to sand and gravel bed channels.

Zeller-Fullerton Equation. Zeller-Fullerton is a multiple regression sediment transport equation for a range of alluvial floodplain conditions. This empirical equation is a computer generated solution of the Meyer-Peter, Muller bedload equation combined with Einstein's suspended load to generate a bed material load [Zeller and Fullerton, 1983]. For a range of bed material from 0.1 mm to 5.0 mm and a gradation coefficient from 1.0 to 4.0, Julien [1995] reported that this equation should be accurate with 10 percent of the combined Meyer-Peter Muller and Einstein equations. The Zeller-Fullerton equation assumes that all sediment sizes are available for transport (no armoring). The original Einstein method is assumed to work best when the bedload constitutes a significant portion of the total load [Yang, 1996].

Summary. The following recommendations pertain to the application of these equations to alluvial fans:

(1) MPM-Smart and MPM-Woo can be used for steep slope, sand bed fans.
(2) Zeller-Fullerton will work on steep slopes and is more appropriate when the bedload is a significant portion of the total load.
(3) Use Ackers-White or Engelund-Hansen equations for subcritical flow in lower sediment transport regime. Using these equations on fan steep slopes may predict excessive sediment transport rates.

(4) The Yang and Karim and Kennedy equations can be used for comparison with the other formulas, but the results may be unrepresentative for steep slope fans.
(5) The Laursen and Toffaleti formulas should be limited to predicting river sediment transport.

It is important to note that in applying these equations, the wash load is not included in the computations. Except for Woo *et al.* [1988] and perhaps MPM-Smart [Smart, 1984], these equations were developed for conventional river flows. Each equation is unique and should be limited to applications that represent the data base that was used to establish the equation (*e.g.*, sand bed or mild slope). The user must research the equation and its applicability to a given project.

Alluvial fan flood studies involving mudflows represent a higher order of complexity because of flow cessation, frontal wave deposits and long runout distances. Sediment flows on alluvial fans range from water flooding to mud floods, mudflows and landslides (Table 4.1). Very viscous, hyperconcentrated sediment flows are generally referred to as mudflows. Mudflows are nonhomogeneous, non-Newtonian, transient flood events whose fluid properties change significantly as they flow across alluvial fans. The mudflow fluid matrix consists of water and fine sediments. At sufficiently high concentrations, fine sediments alter the fluid properties including density, viscosity and yield stress.

Hyperconcentrated sediment flows involve the complex interaction of fluid and sediment processes including turbulence, viscous shear, fluid-sediment particle momentum exchange, and sediment particle collision. Sediment particles can collide, grind, and rotate in their movement past each other. Fine sediment cohesion controls the non-Newtonian behavior of the fluid matrix. This cohesion contributes to the yield stress τ_y which must be exceeded by an applied stress in order to initiate fluid motion. By combining the yield stress τ_y and viscous (η) stress components, the well-known Bingham rheological model is prescribed where v is the velocity in the vertical y profile:

$$\tau = \tau_y + \eta \left(\frac{dv}{dy} \right) \tag{4.1}$$

For large rates of shear such as might occur on steep alluvial fans, turbulent stresses will be generated. In turbulent flow, an additional shear stress component, the dispersive stress, can arise from the collision of sediment particles. Dispersive stress occurs when non-cohesive sediment particles dominate the flow and the percentage of cohesive fine sediment (silts and clays) is small. With increasing high concentrations of fine sediment, fluid turbulence and particle impact will be suppressed and the flow will approach being laminar. Sediment concentration in a given flood event can vary dramatically and as a result viscous and turbulent stresses may alternately dominate, producing flow surges.

The quadratic shear stress model proposed by O'Brien and Julien [1985] describes the continuum of flow regimes from viscous to turbulent/dispersive flow. It is a quadratic rheologic model that includes viscous and turbulent dispersive stresses where C is the turbulent/dispersive coefficient:

$$\tau = \tau_y + \eta \left(\frac{dv}{dy}\right) + C \left(\frac{dv}{dy}\right)^2 \tag{4.2}$$

The first two stress terms in the quadratic model are referred to as Bingham shear stress equation that defines the viscous flow regime. The last term combines the turbulent and dispersive shear stresses that define an inertial flow regime. This term is a function of the square of the velocity gradient. A discussion of these stresses and their role in hyperconcentrated sediment flows can be found in O'Brien *et al.* [1993].

FLO-2D routes mudflows as a fluid continuum by predicting viscous fluid motion as function of sediment concentration. As the sediment concentration changes for a given grid element, dilution effects, mudflow cessation and the remobilization of deposits are simulated. The following empirical relationships can be used to compute viscosity and yield stress:

$$\tau_y = \alpha_2 e^{\beta_2 C_v} \tag{4.3a}$$

$$\eta = \alpha_1 e^{\beta_1 C_v} \tag{4.3b}$$

where α_i and β_i are empirical coefficients defined by laboratory experiment [O'Brien and Julien, 1988]. The viscosity (poises) and yield stress (dynes/cm^2) are shown to be functions of the volumetric sediment concentration Cv of silts, clays and in some cases, fine sands. Very viscous mudflows have high sediment concentrations and correspondingly high yield stresses and may result in laminar flow although laminar flows in nature are extremely rare. Less viscous flows (mud floods) are always turbulent.

For a mudflow event, the average sediment concentration generally ranges between 20 and 35 percent by volume with peak concentrations approaching 50 percent (Table 4.1). Large flood events such as the 100-year flood may contain too much water to produce a viscous mudflow event. Smaller rainfall events such as the 10- or 25-year return period storm may have a greater propensity to create viscous mudflows. Most watersheds with a history of mudflow events develop a sediment supply in the watershed channel such that small storms may generate mudflow surges. Rainfall induced mudflows generally follow a pattern of flood surge response. Initially clear water flows from the basin generated by rainfall-runoff may arrive at the fan apex. This may be followed by a surge or frontal wave of mud and debris (40 to 50 percent concentration by volume). When the peak discharge arrives, the average sediment concentration generally decreases to the range of 30 to 40 percent by volume. On the falling limb of the hydrograph, surges of higher sediment concentration may occur.

When routing the mud flood or mudflow over an alluvial fan or floodplain, the FLO-2D model preserves continuity for both the water and sediment. For every grid element and timestep, the change in the water and sediment volumes and the corresponding change in sediment concentration are computed. At the end of the simulation, the model reports on the amount of water and sediment removed from the study area (outflow) and the amount and location of the water and sediment remaining on the fan or in the channel (storage). The areal extent of mudflow inundation and the maximum flow depths and velocities are a function of the available sediment volume and concentration.

4.4   Important Criteria for Flood Hazard Delineation

When considering the application of a two-dimensional flood model for delineating the flood hazard on an alluvial fan, a conservative approach is to attempt to maximize the area of inundation. This suggests conservative assumptions in flood hydrology and sediment loading analyses. Keeping this in mind, the following recommendations are suggested for developing a 2-D flood routing alluvial fan model:

Flood hydrology volume. The accuracy of predicting the area of inundation is primarily a function of the volume in the flood hydrograph at the fan apex. It is suggested that an appropriate proportion of the work effort and budget (perhaps 50 percent) be devoted to the accurate assessment of the flood hydrograph or rainfall runoff at the fan apex.

Fan Rainfall and Infiltration. Can the fan rainfall contribute to the flooding at the project area? In most cases, the answer would be yes. This could occur through either the rainfall falling directly on the flood water surface or by filling the infiltration storage in advance of the floodwave. The infiltration loss should be calibrated by simulating the fan rainfall without the upper watershed runoff and justifying the percent loss (or percent runoff) on the fan with respect to the upstream basin loss or perhaps other hydrologic studies in the region.

Volume conservation. All flood routing numerical models should report on the volume conservation. The volume conservation (inflow = outflow + storage) should be reviewed for each flood simulation and inflow volume compared with the fan apex hydrology.

Fan apex scenarios. To assess the potential design storm area of inundation, it may necessary to test several fan apex flood scenarios by forcing the flood to opposite sides of the fan using flow path obstructions or by altering the fan topography. The location of the hydraulic control should be identified. If a hydraulic control exists that is higher in elevation than inactive fan areas, it is possible that the inactive areas could be inundated. The topographic apex may be different than the hydrologic apex if the fan channel is incised. This should be represented in the model.

Sediment Bulking. Based on the geology, soils and fan morphology it is possible to surmise the predominant physical processes of sediment transport (flooding or mudflow). If conventional water flooding is presumed as in the case of the desert fans, then bulking the flow by 10% to 20% concentration by volume (Cv) is recommended. Uniformly assigning 10% concentration by volume to the flow would result in a bulking factor $\{BF = 1./1. - Cv)\}$ of 1.11. In this case, simulating a mobile bed might also be considered. If the alluvial fan shows evidence of mudflow and bouldery deposits, then simulating a hyperconcentrated sediment flow is suggested with concentrations ranging from 20% to 50% by volume. The computed mudflow concentrations would be both spatially and temporally variable. Simulating plugged bridges and culverts is an option for mud and debris flows in the FLO-2D model. All alluvial fan flood models should be bulked by at least 10% concentration by volume to account for sediment loading.

Did the simulation run ok? There are number of checks to determine if a numerical model flood simulation produced reasonable results. These include:

(1) Volume conservation;
(2) Numerical surging: Check cell maximum discharges for unreasonable discharge variation;

(3) Maximum channel and floodplain flow depths and velocities (Maximum velocities should be reasonable);

(4) Maximum Froude numbers (Limiting Froude numbers can be assigned in the FLO-2D model); and,

(5) Problem grid elements (FLO-2D identifies floodplain or channel elements that choke the model reduce the computational timesteps).

4.5 Hazard Mapping as a Planning Tool

Around the world, alluvial fan flood hazard mapping is used to regulate development to avoid flood damage and loss of life. In the United States, most of the flood mapping is prepared to establish flood insurance rates. Mapping for the purpose of assessing flood insurance rates rather than for the delineation of the flood hazard to life and property can have negative repercussions. To advance accurate flood hazard mapping, an approach similar to that applied in European countries is presented.

Flood hazard at a specific location is a function of both flood intensity and probability. Flood intensity is determined by the flow depth and velocity. Flood probability is inversely related to flood magnitude; *i.e.*, large flood events occur less frequently. Flood hazard is then defined as a discrete combined function of the event intensity (severity of the event) and return period (frequency). The hazard map criteria presented herein were first proposed for two alluvial fans in Caracas, Venezuela and was later applied to other urbanized fans Garcia *et al.* [2003; 2005]. This approach follows Swiss and Austrian standards that establish three zones to delineate the flood hazard levels as shown in Figure 4.3 [OFEE *et al.*, 1997; Fiebiger, 1997].

The flood hazard map is based on the three color levels to define high (red), medium (orange) and low (yellow) flood hazard levels (Figure 4.3). These map colors translate into specific potential hazard areas as shown in Table 4.2. To define an event's intensity, most methods use a combination of flow depths and velocities. The Austrian method, on the other hand uses the total energy defined as $h + v^2/2g$, where h is the flow depth, v is the velocity and g is the gravitational acceleration [Fiebiger, 1997]. The U.S. Bureau of Reclamation [1988] also identifies hazard as a combination of depth and velocity and differentiates these for adults, cars, and houses as shown in Figures 4.4 through 4.6. The Swiss method defines the intensity in terms of a combination of h and the product of h and v independent of the object subjected to the hazard [OFEE *et al.*, 1997].

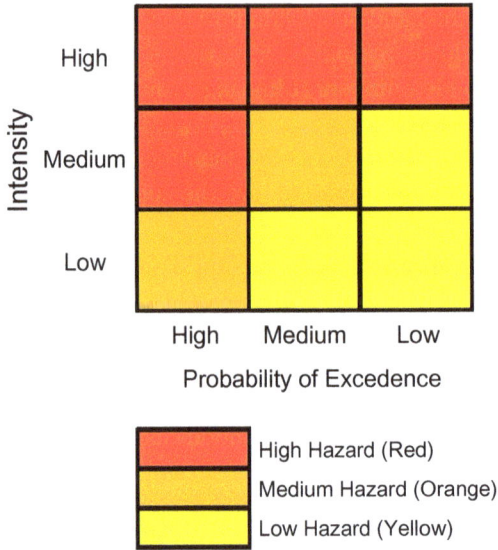

Figure 4.3 Flood hazard levels based on flood frequency and intensity.

Table 4.2 Flood hazard definitions.

Hazard level	Map color	Description
High	Red	Persons are in danger both inside and outside of their houses. Structures are in danger of being destroyed.
Medium	Orange	Persons are in danger outside their houses. Buildings may suffer damage and possible destruction depending on construction characteristics.
Low	Yellow	Danger to persons is low or non-existent. Buildings may suffer little damages, but flooding or sedimentation may affect structure interiors.

Based on researchers such as OFEE *et al.* [1997], Abt [1989] and recent work by Lind *et al.* [2004], all of whom studied the stability of the human body in floods, intensities are defined in terms of the maximum water depth generated throughout the event and the product of the maximum velocity multiplied by the maximum depth. Using the spreadsheet shown in Figure 4.7, the intensity thresholds that differentiates high, medium and

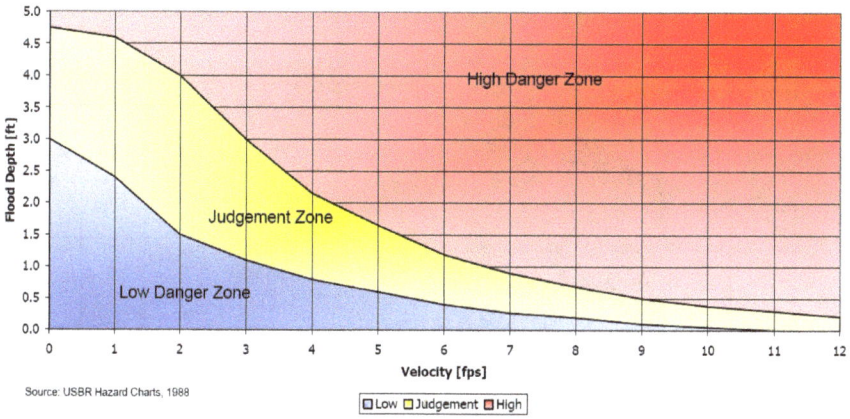

Figure 4.4 Flood hazard for adults (Used with permission from the U.S. Bureau of Reclamation (USBR) [1988]).

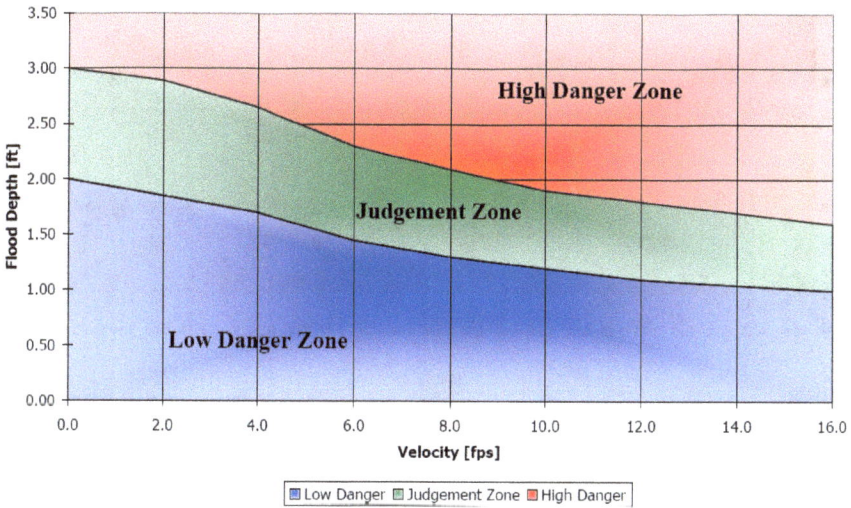

Figure 4.5 Flood hazard for cars (Used with permission from the USBR [1988]).

low levels is compared with the field observed flood impacts during the December 1999 flood in Venezuela [Garcia and Lopez, 2005].

The set of calibrated intensity thresholds presented in Tables 4.3 and 4.4 was obtained by comparing predicted hazard levels using this method with actual affected areas on more than 30 alluvial fans affected by the

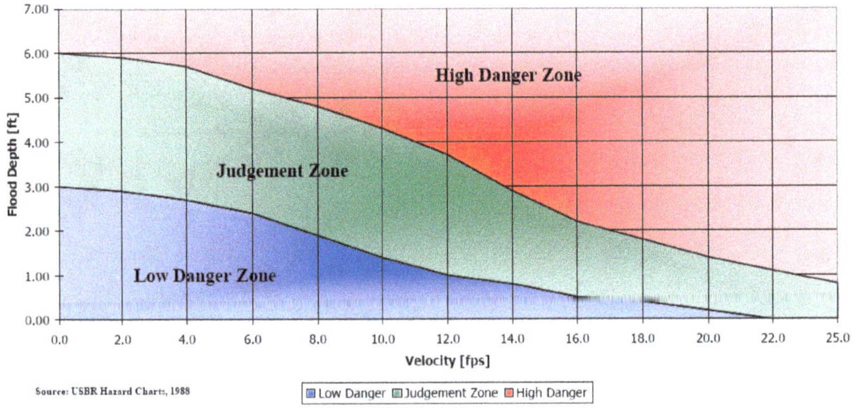

Figure 4.6 Flood hazard for houses (Used with permission from the USBR [1988]).

H																						
5	3	3	3	3	3	3	3	3	3	3	3	3	3	3	3	3	3	3	3	3	3	3	3
4.75	3	3	3	3	3	3	3	3	3	3	3	3	3	3	3	3	3	3	3	3	3	3	3
4.5	3	3	3	3	3	3	3	3	3	3	3	3	3	3	3	3	3	3	3	3	3	3	3
4.25	3	3	3	3	3	3	3	3	3	3	3	3	3	3	3	3	3	3	3	3	3	3	3
4	3	3	3	3	3	3	3	3	3	3	3	3	3	3	3	3	3	3	3	3	3	3	3
3.75	3	3	3	3	3	3	3	3	3	3	3	3	3	3	3	3	3	3	3	3	3	3	3
3.5	3	3	3	3	3	3	3	3	3	3	3	3	3	3	3	3	3	3	3	3	3	3	3
3.25	3	3	3	3	3	3	3	3	3	3	3	3	3	3	3	3	3	3	3	3	3	3	3
3	1	2	2	2	2	2	2	2	2	3	3	3	3	3	3	3	3	3	3	3	3	3	3
2.75	1	2	2	2	2	2	2	2	2	3	3	3	3	3	3	3	3	3	3	3	3	3	3
2.5	1	2	2	2	2	2	2	2	2	2	3	3	3	3	3	3	3	3	3	3	3	3	3
2.25	1	2	2	2	2	2	2	2	2	2	2	3	3	3	3	3	3	3	3	3	3	3	3
2	1	1	2	2	2	2	2	2	2	2	2	2	2	3	3	3	3	3	3	3	3	3	3
1.75	1	1	2	2	2	2	2	2	2	2	2	2	2	2	3	3	3	3	3	3	3	3	3
1.5	1	1	2	2	2	2	2	2	2	2	2	2	2	2	2	2	2	3	3	3	3	3	3
1.25	1	1	2	2	2	2	2	2	2	2	2	2	2	2	2	2	2	2	2	2	3	3	3
1	1	1	1	2	2	2	2	2	2	2	2	2	2	2	2	2	2	2	2	2	2	2	2
0.75	1	1	1	2	2	2	2	2	2	2	2	2	2	2	2	2	2	2	2	2	2	2	2
0.5	1	1	1	1	1	2	2	2	2	2	2	2	2	2	2	2	2	2	2	2	2	2	2
0.25	0	0	0	0	0	0	0	0	0	0	0	0	0	0	0	0	0	0	0	0	0	0	0
0	0	0	0	0	0	0	0	0	0	0	0	0	0	0	0	0	0	0	0	0	0	0	0
VxH	0.00	0.50	1.00	1.50	2.00	2.50	3.00	3.50	4.00	4.50	5.00	5.50	6.00	6.50	7.00	7.50	8.00	8.50	9.00	9.50	10.00	10.50	11.00

Compute Intensity

	ft						ft2/s					
HIGH	1.5	<	h			OR	1.5	<	V x h		
MEDIUM	0.5	<	h	<	1.5	OR	0.5	<	V x h	<	1.5
LOW	0.1	<	h	<	0.5	AND	0.1		v x h	<	0.5

Figure 4.7 Spreadsheet to compute flood and mudflow intensities based on flow depths and velocities.

catastrophic flood and mudflow events of December 1999 in northern Venezuela [Garcia and Lopez, 2005].

In the FLO-2D MAPPER post-processor program, the distinction is made between water flooding and mudflows. Flood intensities are defined in terms of the maximum water depth and the product of the maximum velocity multiplied by the maximum depth. For a specific project, it may be necessary to change the hazard level thresholds. In MAPPER the user

Table 4.3   Definition of water flood intensity.

Flood intensity	Maximum depth h (m)		Product of max depth h times max velocity v (m²/s)
High	$h > 1.5\,m$	OR	$v\,h > 1.5\,m^2/s$
Medium	$0.5\,m < h < 1.5\,m$	OR	$0.5\,m^2/s < v\,h < 1.5\,m^2/s$
Low	$0.1\,m < h < 0.5\,m$	AND	$0.1\,m^2/s < v\,h < 0.5\,m^2/s$

Table 4.4   Definition of mud or debris flow intensity.

Flood intensity	Maximum depth h (m)		Product of max depth h times max velocity v (m²/s)
High	$h > 1.0\ m$	OR	$v\,h > 1.0\ m^2/s$
Medium	$0.2\ m < h < 1.0\ m$	AND	$0.2\ m^2/s < v\,h < 1.0\ m^2/s$
Low	$0.2\ m < h < 1.0\ m$		$v\,h < 0.2\ m^2/s$

can input values for flow depth and velocities that define the intensity thresholds. For the case of water flooding, the flood intensities could be defined by the values in Table 4.3.

Mud flows are more destructive than water floods, thus the mudflow intensity criteria are more conservative (Table 4.4).

The hazard criteria encompass the probability of occurrence of a water or mudflow event for three selected returns periods. This requires a FLO-2D simulation of the three flood frequency events. The model predicts the maximum depths and velocities for each return period flood. For each grid element, the event intensity for a return period flood determines the hazard based on the above criteria. An interpolated shaded color contour plot of grid elements depicts the low, medium and high flood hazards in Figure 4.8.

This method represents a true measure of the flood hazard and is used effectively in other countries. The advantages of this method are clearly apparent. No base flood elevations, flood contours, or first floor building elevation from a FEMA flood insurance rate map are necessary to interpret the flood hazard. From a FEMA DFIRM map perspective, this means that flood elevation alone is not necessarily an indication of flood hazard. The prescribed map method provides an immediate opportunity for the general public to understand whether a given building or neighborhood is within a high flood hazard area. The floodplain manager can decisively plan, regulate and zone based on this flood hazard map and can easily communicate the planning process and hazard to the community.

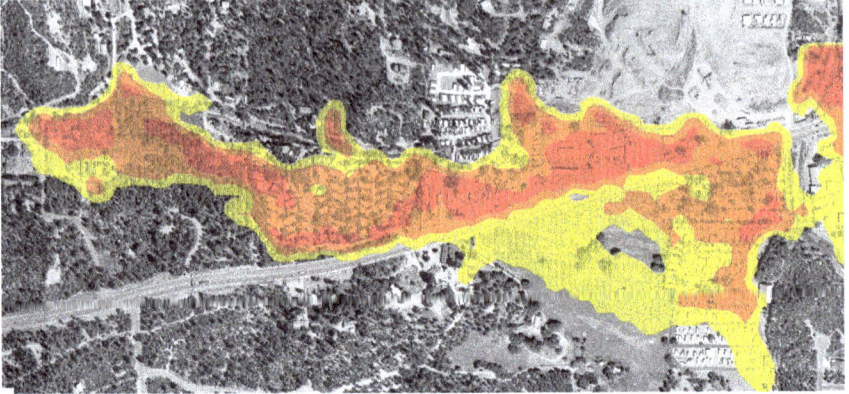

Figure 4.8   An interpolated shaded color contour plot of grid elements depicts the low, medium, and high flood hazards.

4.6   Flood Damage Mapping

An effective measure of flood risk is the cost of flood damage. The Corps of Engineers FDA program is designed to assign flood damage based on damage assessment tables. In the FLO-2D post-processor MAPPER program, the damage assessment using these tables is automated and the total damage cost of the flood simulation can be computed and mapped by structure. Any type of structure or land use such as agricultural crops can be assessed damages. As measure of analyzing flood risk on alluvial fans, flood damage assessment is discussed in Chapter 5.

4.7   Alluvial Fan Mitigation Measures

Mitigation design strategies are discussed for the range of alluvial fan flooding from conventional water flooding to mudflows. Unique designs are required for each project location and potential flood hazard. Mitigation design requires knowledge of the rheological properties of the expected flood or mudflow in a given watershed. All mitigation designs fall into four categories, avoidance, regulatory (zoning), storage or conveyance (or a combination thereof). The first two categories include such measures as elevating on fill, open space or flood easements, and physical removal of structures in the flood path. The last two categories include detention or debris basins, levees and berms, debris fences and deflectors, channelization or channel lining, drop structures, energy dissipation or street alignment.

For water floods and mud floods, the focus should be on flood conveyance off the fan. This means that primary consideration should be given to straight, steep channels. Channel lining, riprap, drop structures, and freeboard are considerations for flood conveyance. Some of the characteristics of mud floods that should be addressed in the mitigation design are:

(1) Sediment bulking;
(2) Roll waves and surging;
(3) Supercritical flow, debris plugging;
(4) Sediment abrasion;
(5) Superelevation; and,
(6) Sediment scour and deposition.

Bed and bank stability in a channel with high Froude numbers are serious concerns and using riprap is not recommended because it could be launched by the flow and contribute to the debris loading. Reducing the slope with drop structures is effective to control the flow energy. The most difficult task in the design of alluvial fan channels is the inlet and outlet transitions to avoid constrictions and debris plugging. Headcutting and undermining protection facilities such as aprons or wing walls are also issues in channel transition reaches.

Mudflow mitigation measures include storage, deflection, flow spreading, and frontal wave dissipation. Mitigation design must consider flow avulsion, debris plugging, and maintenance. Mudflow detention basin opportunities are often limited by steep watershed canyons and the volume estimate for the design event is critical. Deflection and spreading the flow in open space and park areas can be effective mitigation but may incur high maintenance costs. A preferred alternative mitigation measure is an overflow channel with setback levees. The levee is constructed parallel to the channel some appropriate distance from the channel (typically 16 to 33 m) to allow storage of overbank flooding. This concept combines conveyance and storage to enhance flow cessation. The potential for levee erosion and failure must be assessed.

All mitigation facilities on alluvial fans or in the fan watershed require special design considerations including:

(1) Very high velocities;
(2) Impact forces associated with debris and boulders;
(3) Flow runup over mud deposits and structures;

Table 4.5 Freeboard and factor of safety recommendations.

Type of flooding	Freeboard (ft)	Impact factor of safety
Shallow water flooding <1 ft	1	1.1
Moderate water flooding <3 ft	1	1.2
Moderate water flooding <3 ft; debris, boulders <1 ft	1	1.2
Mud floods, debris flow <3 ft, surging, debris, sediment deposition, boulders <1 ft	2	1.25
Mudflow, debris flow <3 ft, surging, debris, sediment deposition, waves, boulders >1 ft	3	1.4
Mudflow, debris flow >3 ft; surging, waves, boulders >3 ft	3–5	1.5

(4) Static pressures on structures;
(5) Extra freeboard requirements (Table 4.5);
(6) Sediment bulking; and,
(7) Cleanup and maintenance access.

Both detention basin and conveyance facilities require evaluation of the volume and peak discharge of the design flood event. The generally accepted method is to bulk the 100-year hydrograph for the potential average sediment concentration by volume. A conservative approach is to use a concentration by volume of 50 percent resulting in a bulking factor of 2.

References

Abt, S.R., Wittier, R.J., Taylor, A. and Love, D.J. (1989). "Human Stability in a high flood hazard zone." AWRA, Water Julien, P.Y., 1995. Erosion and Sedimentation. Cambridge University Press, New York, N.Y. Resources Bulletin, V. 25, No. 4, pp. 881–890.

Ackers, P. and White, W.R. (1973). "Sediment transport: New approach and analysis." *J. of Hyd., ASCE*, 99(HY11), 2041–2060.

Bagnold, R.A. (1954). "Experiments on a gravity-free dispersion of large solid spheres in a Newtonian fluid under shear." Proceedings Royal Society of London, Series A, 225, 49–63.

Dawdy, D.R. (1979). "Flood frequency estimates on alluvial fans," *Journal of the Hydraulics Division*, 105(HY11), 1407–1412.

DeLeon, A.A. and Jeppson, R.W. (1982). "Hydraulics and numerical solutions of steady-state but spatially varied debris flow." Hydraulics and Hydrology Series, UWRL/H-82/03, Utah Water Research Laboratory, Utah State Univ., Logan, Utah.

Engelund, F. and Hansen, E. (1967). "A monograph on sediment transport in alluvial streams." Teknisk Forlag, Copenhagen.

Federal Emergency Management Agency (2002). "Guidelines and specifications for flood hazards mapping partners, Appendix G, Guidance for alluvial fans flooding analyses and mapping." Washington, D.C. http://www.fema.gov/mit/ft_alfan.htm.

Fiebiger, G. (1997). "Hazard mapping in Austria." *Journal of Torrent, Avalanche, Landslide and Rockfall Engineering*, 134, 61.

French, R. (1991). "Preferred directions on flow on alluvial fans." *ASCE Journal of Hydraulic Engineering*, 118(7), 1002–1013.

Garcia, R., López, J.L., Noya, M., Bello, M.E., Bello, M.T., González, N., Paredes, G., Vivas, M.I. and O'Brien, J.S. (2003). "Hazard mapping for debris flow events in the alluvial fans of northern Venezuela." Third International Conference on Debris-Flow Hazards Mitigation: Mechanics, Prediction and Assessment. Davos, Switzerland. September 10–12.

Garcia, R. and Lopez, J.L. (2005). "Debris flows of December 1999 in Venezuela." Chapter 20th of Debris-flow Hazards and Related Phenomena. Jakob, Matthias, Hungr, Oldrich Eds. Springer Verlag Praxis, Berlin.

Graf, W.H. (1971). Hydraulics of Sediment Transport. McGraw-Hill, New York, N.Y.

Julien, P.Y. (1995). Erosion and Sedimentation. Cambridge Univ. Press, New York, N.Y.

Karim, F. (1998). "Bed material discharge prediction for nonuniform bed sediments." *ASCE Journal of Hydraulics*, 124(6), 597–604.

Laursen, E.M. (1958). "The total sediment load of streams." *ASCE Journal of the Hydraulics Div.*, 84, 1530–1536.

Lind, N., Hartford, D. and Assaf, H. (2004). "Hydrodynamic models of human stability in a flood." *Journal of the American Water Resources Association*, 40(1), 89–96.

Lowe, D.R. (1976). "Grain flow and grain deposits." *Journal of Sedimentary Petrology*, 46, 188–199.

Mifflin, E.R. (1988). "Design depths and velocities on alluvial fans." Proc. of the ASCE National Conference on Hydraulic Engineering, ASCE, New York.

Mifflin, E.R. (1990). "Considering entrenched channels when modeling alluvial fan flooding." Proc. of the ASCE Intl. Symposium of Hydraulics and Hydrology of Arid Lands, ASCE, New York, pp. 28–33.

Mussetter, R.A., Lagasse, P.F., Harvey, M.D. and Anderson, C.A. (1994). "Sediment and Erosion design guide." Prepared for the Albuquerque Metropolitan Arroyo Flood Control Authority by Resource Consultants and Engineers, Inc., Fort Collins, CO.

National Research Council (1983). "An evaluation of flood-level prediction using computer-based models of alluvial rivers." Committee on Hydrodynamic

Computer Models for Flood Insurance Studies, Advisory Board on the Built Environment, Commission on Engineering and Technical Systems, National Academy Press, Washington, D.C.

O'Brien, J.S. and Julien, P.Y. (1985). "Physical processes of hyperconcentrated sediment flows." Proc. of the ASCE Specialty Conf. on the Delineation of Landslides, Floods, and Debris Flow Hazards in Utah, Utah Water Research Laboratory, Series UWRL/g-85/03, 260–279.

O'Brien, J.S. (1986). "Physical processes, Rheology and modeling of mudflows." Ph.D. Dissertation, Colorado State Univ., Fort Collins, Colorado.

O'Brien, J.S. and Julien, P.Y. (1988). "Laboratory analysis of mudflow properties." *J. of Hyd. Eng., ASCE*, 114(8), 877–887.

O'Brien, J.S., Julien, P.Y. and Fullerton, W.T. (1993). "Two-dimensional water flood and mudflow simulation." *J. of Hyd. Eng., ASCE*, 119(2), 244–259.

OFEE, OFAT, ODEFP (Switzerland) Ed., (1997). "Prise en compte des dangers dus aux crues dans le cadre des activités de l'aménagement du territoire." Office fédéral de l'économie des aux (OFEE), Office fédéral de l'aménagement du territoire (OFAT), Office fédéral de l'environnent, des forets et du paysage (OFEFP), Bienne.

Savage, S.B. (1979). "Gravity flow of cohensionless granular materials in chutes and channels." *Journal of Fluid Mechanics*, 92(Part 1), 53–96.

Savage, S.B. and McKeown, S. (1983). "Shear stresses developed during rapid shear of concentrated suspensions of large spherical particles between concentric cylinders." *Journal of Fluid Mechanics*, 127, 453–472.

Schamber, D.R. and MacArthur, R.C. (1985). "One-dimensional model for mudflows." Proc. of the Conference Hydraulics and Hydrology in the Small Computer Age, ASCE, Orlando, Florida, p. 1334f.

Smart, G.M. (1984). "Sediment transport formula for steep channels." *ASCE Journal of Hydraulic Engineering*, 110, 267–275.

Takahashi, T. (1978). "Mechanical characteristics of debris flow." *Journal of the Hydraulics Division, ASCE*, 104, 1153–1169.

Takahashi, T. (1980). "Debris flow on prismatic open channel." *Journal of Hydraulics Divisions, ASCE*, 106, 381–396.

Takahashi, T. and Tsujimoto, H. (1985). "Delineation of the debris flow hazardous zone by a numerical simulation method." Proc. of the Intl. Symp. on Erosion, Debris Flow and Disaster Prevention, Tsukuba, Japan, 457–462.

Takahashi, T. and Nakagawa, H. (1989). "Debris flow hazard zone mapping." Proc. of the Japan — China (Taipai) Joint Seminar on Natural Hazard Mitigation, Kyoto, Japan, 363–372.

Toffaleti, F.B. (1969). "Definitive computations of sand discharge in rivers." *ASCE J. of the Hydraulics Div.*, 95, 225–246.

U.S. Army Corps of Engineers (1988). "Mud flow modeling, one- and two-dimensional, Davis county, Utah." Draft report for the Omaha District, Omaha, NE, October.

USBR (1988). Downstream Hazard Classification Guidelines. ACER Technical Memorandum No. 11. Woo, H.S. Julien, P.Y. and Richardson, E.V. (1988).

"Suspension of large concentrations of sand." *J. Hyd. Eng., ASCE*, 114(8), 888–898.

Woo, H.S., Julien, P.Y. and Richardson, E.V. (1988). "Suspension of large concentrations of sands." *Journal of Hydraulic Engineering*, ASCE 114(8), 888–898.

Yang, C.T. (1973). "Incipient motion and sediment transport." *J. of Hyd Div. ASCE*, 99(HY10), 1679–1704.

Yang, C.T. (1996). "Sediment transport, theory and practice." McGraw-Hill, New York, N.Y.

Yano, K. and Daido, A. (1965). "Fundamental study on mudflow." Disaster Prevention Research Institute, Kyoto Univ., Kyoto, Japan, Annuals, No. 7, pp. 340–347.

Zeller, M. E. and Fullerton, W.T. (1983). "A theoretically derived sediment transport equation for sand-bed channels in Arid regions." Proceedings of the D. B. Simons Symposium on Erosion and Sedimentation, R.M. Li and P.F. Lagasse, eds., Colorado State University and ASCE.

Chapter 5

Flood Hazard Mapping Versus Flood Risk Analysis

Jimmy S. O'Brien

FLO-2D Software, Inc., P.O. Box 66
102 County Road 2315, Nutrioso, Arizona 85932
jim@flo-2d.com

Reinaldo Garcia

Applied Research Center, Florida International University
10555 West Flagler Street EC 2100, Miami, Florida 33174
reinaldo@flo-2d.com

This chapter outlines methods to assess flood hazard and flood risk on alluvial fans. Concepts of flood hazard and risk are discussed as well as uncertainties in predicting floods. Both stochastic and deterministic predictive methods are presented. As an example of a practical stochastic method, a random walk algorithm to compute potential flow paths in the context of a Monte Carlo procedure is outlined. In addition, a conceptual model to link stochastic and deterministic approaches is proposed. Finally, as an example of risk map delineation, the FLO-2D damage assessment procedure is presented that computes flood costs based on damage-depth tables and deterministic model results.

5.1 Risk and Uncertainty of Alluvial Fan Flooding

Alluvial fan flooding has flow path uncertainty through the combination of variable topography, sediment load and possible obstructions. Unconfined flows, through floodwave attenuation diminish the hydraulic intensity of the flood hazard but they increase the potential area of inundation. Conversely channelized flows and channel incision increase the flood intensity

89

and the frequency of the flooding given that the incised channel remains locked in place for a period of time. Flow path uncertainty is linked to the sediment supply and possible channel avulsions. The hydrology of flood frequency plays a role in stimulating the fan avulsions by charging the upstream watershed water courses with a buildup of sediment through a sequence of smaller runoff events.

In this chapter, flood hazard delineation and flood risk on alluvial fans due to the potential flow path uncertainty is analyzed. Flood hazard and risk assessment are critical components for land use planning, floodplain management, mitigation design, emergency preparedness, and disaster response. Flood risk and flow path uncertainty by necessity invoke consideration of stochastic methods. The role of deterministic models in flood hazard delineation, as discussed in the previous chapter, are critical for the flood hazard delineation but are often viewed as not encompassing the nature of flood risk and fan geomorphic evolution. Some concepts of flood hazard and flood risk need to be defined to begin this discussion.

5.1.1   Concepts of flood hazard and flood risk: Hazard ≠ risk

Hazard as defined by the UN-ISDR [2004] is "...a potentially damaging physical event, phenomenon or human activity that may cause the loss of life or injury, property damage, social and economic disruption or environmental degradation." Flood hazard is defined as exposure to flooding. It is the probability of a design flood intensity occurring at given fan or floodplain location and can be represented as the potential flood inundation on a map. The design flood event has been ubiquitously selected for most flood studies as the 100-year flood (or a flood with 1.0 percent probability of occurrence in any given year) and was adopted as a standard for floodplain management by federal, state and local agencies as an acceptable level of flood hazard. On alluvial fans, the flood hazard areas are defined as AO zones on flood insurance rate maps, corresponding to the 100-year shallow flooding (usually referred to as sheet flow on variable topography). The fan is subdivided into similar areas of flow depth (or specific energy) and velocity [Zhao and Mays, 1996].

Design flood criteria may be changing in favor of larger magnitude flood events. For example, the State of California, Department of Water Resources' Central Valley Floodplain Evaluation and Delineation Project will require floodplain delineation maps for the 100-, 200-, and 500-year flood events. California passed legislation effective January 1, 2008 that prohibits cities and counties in the Sacramento and San Joaquin River Valleys from permitting urban development projects unless the project has 200-year flood protection [DWR, 2008].

Often when people discuss risk, then mean hazard. Risk is defined by the UN-ISDR [2004] as "...the probability of harmful consequences, or expected losses (deaths, injuries, property, livelihoods, economic activity disrupted or environment damaged) resulting from interactions between natural or human-induced hazards and vulnerable conditions". The goal of most alluvial fan flood studies is to map the flood hazard considering uncertainties in physical conditions such as flow roughness, topographic elevations, obstructions, and high sediment loads, all of which contribute to potentially variable flow paths. Mapping the flood hazard with a vulnerability assessment will lead to an evaluation of flood risk. Risk can by expressed by the equation:

$$R\,(\text{Risk}) = H\,(\text{Hazards}) \times V\,(\text{Vulnerability}) \qquad (5.1)$$

where the hazard H is the probability (annual) of a point location on the alluvial fan being inundated and V is probability of expected losses (destruction of buildings) on the fan. Probability is a measure of risk. A quantitative risk analyses is a function of the hazard, the elements at risk and the vulnerability according to the Committee on Risk Assessment of the Working Group on Landslides of the International Union of Geological Sciences [IUGS, 1997]. The vulnerability of the persons or objects which constitute the 'elements at risk' in a flood hazard area is weighted according flood frequency [Plate, 2002]. Calvo and Savi [2008] determined the expected annual risk for buildings located on a debris flow fan by integrating the vulnerability exceedance probability curve. The total risk must then be evaluated by multiplying the specific annual risk by the cost of the damaged properties.

Concepts of risk, vulnerability and hazard are applied more universally with GIS mapping to understand urban population dynamics and regulations, establish emergency management and plan disaster relief [Maantay and Maroko, 2008]. The risk analysis process can yield both hazard and risk maps which can be plotted as color shaded contours on aerial photos (Figure 5.1). Risk analysis can be conducted for the hazard flood zone on alluvial fans. The challenge is to assess the probability that a given point on the fan will be located in the hazard flood zone. This is a difficult proposition because of the lack of objective measures of acceptable risk,

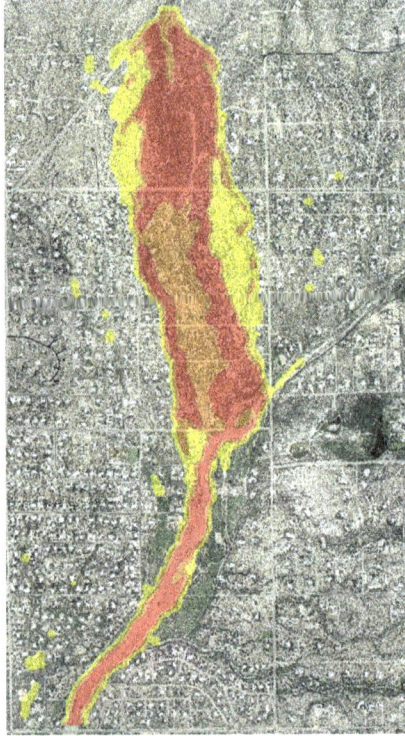

Figure 5.1   Flood hazard map for Soldier Canyon alluvial fan, Tucson, Arizona.

scarcity of alluvial fan flood data and unknown flood parameter probability distributions. Typically probability distribution functions for the uncertain variables are selected and fitted to the available data. However, it is unlikely that the extreme rare events will be generated from the same flow regime that constitutes the data base [Lind *et al.*, 2008].

5.2   Stochastic versus Deterministic Flood Hazard Assessment

There are two types of models that are used for the spatial analysis of flood hazards; deterministic models and stochastic process models. Deterministic models are based on physical relationships and predict the flooding based entirely on the input conditions (existing fan conditions). For deterministic models no experimental or sample data is required. Stochastic models (or probabilistic models) evaluate a flood hazard as a random process. The stochastic model uses random variables that have assumed or known probability distributions and will generate repeated output that may not be the same, but will follow statistical patterns. Deterministic models always produce the same output from the same set of input data.

Flooding on alluvial fans is often described as a stochastic process because the number of variables affecting the flood depth may exceed the capabilities of the physically based model to predict them. Stochastic models can help define the flooding probability on an unconfined surface by treating the physical system as an indexed collection of random variables. The stochastic process can be defined as a Markovian property if the probability of the random variables of being in any future state, given any past and present states depends solely on the present state [Price, 1974].

Deterministic models such as those discussed in the previous chapter predict flow depth and velocity as a function of the inflow hydrograph. For example, the FLO-2D flood routing model predicts flow depth and velocity by distributing flood volumes over a system of square cells that represent the topographic flow domain. With this model, flooding can be predicted as a function of flood frequency, duration and magnitude and results can be accurately mapped for a given set of physical conditions. During a flood event, topography, bedforms, obstructions and flow roughness can be variable and have uncertainty even when the physical models have been calibrated.

A stochastic model application to delineate fan flood hazard can display flood risk as a probability of inundation, but the stochastic model is limited by its inability to predict flood intensity or vulnerability. Deterministic models can predict flood hydraulics for a given set of conditions to define the flood hazard intensity, but cannot directly establish the probability of flood inundation for variable future conditions. The best of both worlds is to conduct a flood hazard study that identifies a range of potential future conditions through a stochastic process and then predicts the hydraulic conditions with a deterministic model.

5.3   Stochastic Methods for Fan Flood Hazards

Flood hazard analyses are primarily driven and controlled by the project hydrology which establishes the flood volume and the overall area of inundation while hydraulic routing processes play a secondary role in distributing the flood to define the inundation details. Flood hazard uncertainty and risk is embodied in variables that define the flood volume or the flow hydraulics (velocity and depth). There are several methods for conducting uncertainty analyses for hydrologic or hydraulic models including first-order, second moment approximation methods, probabilistic point methods, Monte Carlo simulations, and integral transformation methods [Zhao and Mays, 1996]. The Monte Carlo simulations method will be discussed further. From a practical standpoint, the objective is to link together the stochastic and deterministic methods to combine reliable flood hazard mapping with the associated flood risk.

5.3.1   *Monte Carlo simulations*

Hydrologic and hydraulic models can be applied with parameters that vary randomly but follow assumed probability distributions that can be estimated from a Monte Carlo simulation. A Monte Carlo approach to flood prediction requires repeated simulations using prescribed values of the random variables generated from the variable probability distribution. The results of the Monte Carlo simulation are statistically significant if a sample of the results reflect the observed results. Generating the variable probability distributions is the critical task for successful application of the Monte Carlo method to alluvial fan analyses.

Applying a Monte Carlo method to alluvial fans has been undertaken in several studies [Calvo and Savi, 2008; Wichmann and Becht, 2003; and Price, 1974]. Typically the Monte Carlo method is applied to watershed hydrology in terms of rainfall sequence to establish return period flooding or debris flows [Calvo and Savi, 2008]. Unfortunately applying Monte Carlo methods to fan flooding is problematic because the hydraulic models (particularly flood routing models) are computer runtime intensive often taking on the order of hour(s) to complete one simulation. The required number of flood routing simulations, perhaps 10,000 computer simulations or more, render the application of the Monte Carlo method in this manner impractical.

A possible approach to applying a Monte Carlo method that does not require repeated simulations of a flood routing model is the random walk method [Price, 1974; Wichmann and Becht, 2003]. The random walk method employs a random number generator to estimate parameters from user defined probability distributions. The uncertainty parameters for alluvial fans include gradient, roughness, and possible flow obstruction to calculate a preferred direction on a discretized grid system. Starting with each inflow grid element, a single pass would constitute a primary flow path over the alluvial fan. By summing the number of times a grid element is selected in the preferred process path out of total number flow path computations (thousands), the probability that a grid element would be inundated for a given flood event would be determined. A random walk model would fit well with the FLO-2D model grid system and it is discussed in more detail in a following section. Combining the deterministic model to predict depth and velocity with the stochastic random walk method to determine probability risk would result in the complete definition of the alluvial fan flood risk including active and inactive portions of the fan.

5.3.2 *Probability distributions representing physical fan parameters*

Deterministic models can predict unsteady and unconfined flood hydraulics in two dimensions based on existing conditions. Due to variability in future fan conditions, such as topography, roughness, and avulsions, predicted flow depths and velocities have a potential uncertainty. As was previously mentioned, examining these hydraulic variable uncertainties through a Monte Carlo method would require a large number of flood simulations that would be infeasible. Using a random walk stochastic model based on probability distributions, the potential inundation probability could be assessed. The final product would then be a map indicating the potential inundation probability (on a grid element basis) overlaid with the predicted maximum flow depths for a given return period flood.

The underlying foundation for the random walk model or any Monte Carlo method is the probability distributions for the selected variables. Flood variable probability distribution functions are discussed by a number of authors [French, 1992; Calvo and Savi, 2008; Zhao and Mays, 1996; van Mamen and Brinkhuis, 2005; Price, 1974; Heggen, 1994; Romanowicz, R. and K. Beven, 2003]. The general form of the probability distribution for some random variable H at a discrete location on the alluvial fan is:

$$P_{(H|X)} = \int_{X_0}^{\infty} P_{(H|X)} f_x(x) dx \qquad (5.2)$$

where x is a second random variable (*e.g.*, discharge), $P_{(H|X)}$ is the probability that the point on the fan will be inundated, and $f_x(x)$ is the probability density function. For an alluvial fan hydraulic model, the uncertainty parameters that might be considered for probability density functions for a Monte Carlo simulation include:

(1) Topography or slope;
(2) Roughness value;
(3) Momentum change;
(4) Obstructions (levees, buildings);
(5) Flow path preference or persistence (channels, streets, distributary channels);
(6) Sediment concentration (optional); and,
(7) Scour and deposition (channel avulsion).

These parameters are used to compute the dependent variables depth and velocity. Consideration of sediment properties adds another layer of complexity and would include uncertainties related to size fraction, volume concentration, viscosity, yield stress, and percent clay fraction. If we ignore the sediment component for the moment and instead perhaps use an avulsion function [Heggen, 1994], then there are six parameters that would be determined by probability functions. The probability function for the avulsion (av) or obstruction (ob) may be simple such that if there is no opportunity for avulsion or obstruction (av = 0. or ob = 0.), then the probability $P = 1.0$ for the given flow direction. Otherwise, for an occurrence of avulsion or obstruction (av = 1. or ob = 1.), then there is no probability for flow in that direction $P = 0.0$. The other parameters would have more complex probability functions. For example, Price [1974] computes the probability for the slope (s) as:

$$P = 0.25 + 0.75\,s \qquad (5.3)$$

such that if the slope is 0.0, there is a 25% probability in the flow choosing one of the four paths in a uniform grid pattern (this would correspond to a 0.125 probability for the 8-direction FLO-2D model). For flow roughness, a normal distribution about a mean value on different portions of the fan might be appropriate. The distribution function could be created from the frequency of the spatially variable grid element n-values on different portions of the fan using the FLO-2D grid system. Topography (slope), roughness and momentum can be assigned probabilities based on a random estimate within a given range of the prescribed probability function. The probability for the variables can be accomplished with a Monte Carlo sampling technique. The generation of uniformly distributed random numbers between 0 and 1 can be used to assign a probability for potential scour and deposition, variation in flow roughness and changes in flow path direction. Obstructions, flow path preference and avulsion are assigned probabilities on the basis of a switch (on or off).

Working with probability distribution functions requires some mathematical and numerical sophistication. At a minimum, it is necessary to understand how to convert a discrete number of field or experimental observations or an assumed range of values into a full probability distribution.

5.3.3 *Random walk algorithm to determine flow paths*

A method to determine flow paths using a random walk algorithm similar to that developed by Gamma [2000] is proposed. A FLO-2D grid system consists of square grid elements. Each element (cell) has a representative bed elevation generally obtained by interpolation of irregularly located elevation points of a Digital Elevation Model (DEM). Starting with a given grid element, all lower elevation neighbor elements are potential flow path cells. Using slope threshold and divergent flow parameters, the method allows the potential flow path (random walk) to choose directions that have a lower elevation. The flow will follow the steepest descent if the slope to the neighbor is greater than the slope threshold. This results in a single flow direction algorithm, analogous to the well known D8 algorithm [Jenson and Domingue, 1988]. The probability for each grid element to be selected from the set of potential path elements is given by:

$$
p_i = \begin{cases} \displaystyle\sum_j \frac{\tan\beta_i\,p}{\tan\beta_j}, & \text{if } k \in N \\[2mm] \displaystyle\sum_j \frac{\tan\beta_i}{\tan\beta_j}, & \text{if } k \in N \end{cases} \tag{5.4}
$$

where N is the set of potential flow directions for the given element, β_i is the bed slope to neighbor element i, p is a persistence factor and k is the previous flow direction. The calculated transition probabilities are scaled to cumulative values in the [0,1] interval and a random number generator selects one flow path element from the N set.

For each starting point of a flow path, random walks are calculated, each one resulting in a slightly different path. For example, Figure 5.2 shows an idealized terrain where each block represents a grid element. Flow paths A and B are of two of the potential paths from the fan apex that would represent inundation of different elements for each random walk computation.

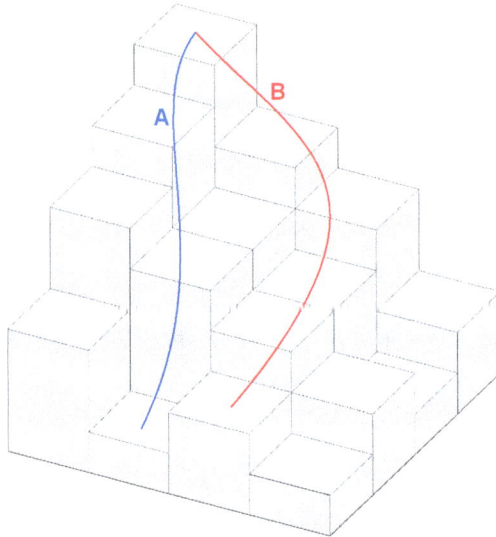

Figure 5.2   Flow paths A and B represent two of probable flow paths from the fan apex.

If a sufficiently large number of paths are calculated, essentially the entire potential flow area will be inundated. By counting how often a grid element is intersected by the flow path, the relative measure of process intensity can be computed. This intensity represents an indication of which grid elements are more likely to be affected by the flow.

This approach has a number of interesting properties. For example, the slope threshold allows the model to adjust to different topographies. In steep areas with slopes close to the threshold, only steep neighbors are permitted in a random walk path in addition to the path of steepest descent. In flat areas, almost all lower elevation neighbors are possible flow path cells. Abrupt changes in flow direction are minimized by applying a higher persistence factor. Also the steepest descent path tends to have a higher frequency as the probabilities are weighed by slope. For more details refer to Gamma [2000] and Wichmann and Becht [2003].

5.3.4   *Alluvial fan flood probability — creating the linkage between the stochastic model and the deterministic model*

The modeler has the responsibility of developing both the stochastic method and the probability distribution functions. There are no commercially available models to perform this function. The California Department of

Water Resources (DWR) may develop a stochastic tool for mapping alluvial fan hazards within the boundaries of its Central Valley Floodplain Evaluation and Delineation project. The FLO-2D model was selected by DWR as the two-dimensional flood routing model for this project for both rivers and alluvial fans. It is proposed that the random walk stochastic method be applied with Monte Carlo computations of random values of the probability distributions of the selected variables. The inundation probability will then be established for the grid elements through a large number of random walk path simulations (*e.g.*, 10,000 runs). The procedure to conduct the probability analysis is to:

(1) Identify the primary variables or processes;
(2) Assign probability distribution to the variables;
(3) Perform the Monte Carlo random number calculation;
(4) Compute the grid element inundation probabilities to generate the random walk path (path with the highest probability for that simulation);
(5) Conduct the random walk path simulations; and,
(6) Calculate the inundation probability of the each grid element (based on the number of hits that each grid element received out of total number of path simulations).

Steps 3 through 6 would be performed as a pre-processor program in the FLO-2D model system. The final flood hazard map would be a combined overlaid map of:

(1) Predicted maximum flow depths for the existing conditions; and,
(2) Inundation probability or flood risk.

It would also be possible to map the predicted maximum depths associated with the average, maximum and minimum assigned variables computed from the random number generation of the probability distributions. The results would be interpreted as a predicted range of maximum flow depths, velocities, impact pressure, *etc.* for each grid element which would also have an assigned probability of inundation.

5.3.5 *Evolution of the alluvial fan — modeling future conditions*

As suggested in the previous section, by selecting or by filtering the probability distribution functions for prescribed variables, a range of potential future conditions could be simulated. For example, the topography (or slope) probability distribution (normal, Poisson, *etc.*) could range from $+1$

to -1 meter. This resulting series of flood scenarios would reflect deposition (or scour) on the elements near the fan apex. Similar distributions could be selected for n-values or other variables.

It follows that this scenario approach could be applied using a recent advancement in numerical modeling called ensemble prediction systems (EPS). Initially ensemble prediction methods were developed by meteorologists to address limitations of deterministic weather forecasting in light of the uncertainty (chaos) of weather variables [Demeritt *et al.*, 2007]. Whereas traditional deterministic models produce a single prediction based on existing conditions, an EPS model will generate an ensemble of predictions. An ensemble suite of tests could have varying initial or boundary conditions or varying model parameters to reflect the project uncertainty. Preparing and running a suite of model simulations for each possible combination of variables can be computer intensive. Data file editing and compiling results for detailed models with a large grid system of small elements may be difficult or impossible. A coarser grid system with less detail may be necessary for EPS models.

A suite of predictions, rather than a single deterministic model simulation, provides a way to improve the linkage between the stochastic product of flood probability and the deterministic model and will communicate the uncertainty in the hazard mapping. The EPS results may also be packaged as a probability distribution. It would provide an explicit method of displaying flood risk for insurance purposes.

Improving flood uncertainty predictions is welcomed by floodplain managers charged with the responsibility to conduct cost-loss analyses. With flood inundation probability maps, floodplain managers can balance flood hazard regulations, zoning and mitigation with accurate flood insurance rates based on actual flood risk. Eventually EPS will become the state-of-the-art technique for floodplain and alluvial fan mapping. It is currently in daily operational use by weather service agencies around the world for accurate weather forecasting. Faster computing resources is a key for EPS application to flood risk.

5.4   Integrating Alluvial Fan Flood Hazard Mapping and Damage Assessment

A true measure of flood risk is the actual damage cost resulting from the exposure of *vulnerable* structures to a given return period flood *hazard*. This is represented by the linkage of the flood hazard map and damage cost assessment technique. Damage assessment requires significant field effort to evaluate the potential cost of inundation as function of flood depth. Similar to the Corps of Engineers Flood Damage Assessment (FDA) program, the following method for estimating the total and individual structure flood damage has been automated in the FLO-2D Mapper post-processor program. This tool can be used to estimate the damage cost due to flood inundation for any type of structure or land use (*e.g.*, agricultural crops). To apply this method in the Mapper program, the following data must be available:

(1) A polygon shape file where each polygon represents a structure or field.
(2) A table file containing damage cost data as function of flood depth for each building type in the polygon shape file. The file will have a code that will correspond to a shape file polygon and cost data for damage per foot or meter of flow depth.

The procedure used to calculate damage cost is as follows:

(1) Read maximum flood depth in each grid element resulting from an application of the FLO-2D model.
(2) Import building polygon shape file.
(3) Import cost table ($US or other monies associated with the flow depth for each structure, field, *etc.*).
(4) Determine building (or field) polygon intersections with the FLO-2D grid elements.
(5) For the building (or field) polygons, calculate damage ($US) using the following weighted average formula:

$$COST_B = \left(\frac{\sum A_i (Cost_i)}{A_B} \right) \tag{5.5}$$

where $COST_B$ is the estimated cost caused by the flooding for building B, A_i is the subarea of intersection between the building and the grid element i. $Cost_i$ is the estimated cost associated with the area flooded by grid element i, and A_B is the area of building B.

When the algorithm is complete, each building or field will have an estimated flood damage cost estimate. Figure 5.3 shows the FLO-2D predicted maximum flood depth and the imported building polygon shape file.

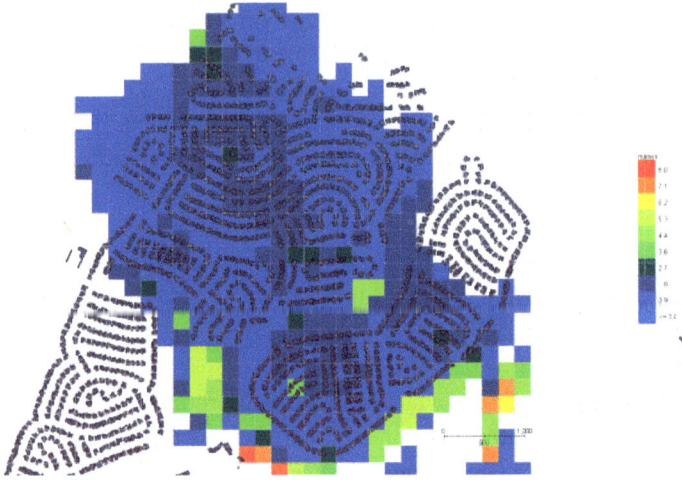

Figure 5.3   FLO-2D predicted grid element maximum flow depth overlaid with the building polygon shape file as shown in a Mapper display.

The Mapper damage assessment table (Figure 5.4) is similar to that used by the USCOE FDA program and can easily accommodate other depth-damage curves similar to the Federal Insurance Administration's (FIA) *credibility weighted* curves used in the HAZUS model [Scawthorn *et al.*, 2006]. For each building identified by a polygon ID, the table provides damage cost for up to 10 user specified depths (D1-D10). When the damages are computed, the Mapper program will compute all the damages related to the flow depth according to the portion of the building area covered by the grid element. It will also create a GIS polygon shape file containing the damage cost for each building intercepted by the flooding area and the total damage will be displayed (Figure 5.5).

The Mapper will display the damage for each building by color assignment (Figure 5.6) and will display the actual damage cost value (Figure 5.7).

The damage cost assessment table as designed by the USCOE was established for conventional floodplain inundation. The damage cost tables have to reflect the alluvial fan flood, mudflow or debris damages associated with high velocities or impact pressures. The damage tables also have to reflect the types of structures that may be designed to withstand debris flows including raised first floors, masonry walls, and lack of windows and doors

Flooding Damage Estimation

Buildings shape file: HOUSES.SHP

Building Id field: HOUSES_ID

Import Damage Table... C:\FLO-2D\2006.01\Projects\Damage_assessment_example\DamageCode_08.tbl

Id	D1	D2	D3	D4	D5	D6	D7	D8	D9	D10
1	3000	6000	9000	12000	15000	18000	21000	24000	27000	30000
2	3000	6000	9000	12000	15000	18000	21000	24000	27000	30000
3	3000	6000	9000	12000	15000	18000	21000	24000	27000	30000
4	3000	6000	9000	12000	15000	18000	21000	24000	27000	30000
5	3000	6000	9000	12000	15000	18000	21000	24000	27000	30000
6	3000	6000	9000	12000	15000	18000	21000	24000	27000	30000
7	3000	6000	9000	12000	15000	18000	21000	24000	27000	30000
8	3000	6000	9000	12000	15000	18000	21000	24000	27000	30000
9	3000	6000	9000	12000	15000	18000	21000	24000	27000	30000
10	3000	6000	9000	12000	15000	18000	21000	24000	27000	30000
11	3000	6000	9000	12000	15000	18000	21000	24000	27000	30000
12	3000	6000	9000	12000	15000	18000	21000	24000	27000	30000

Compute Damage... Cancel

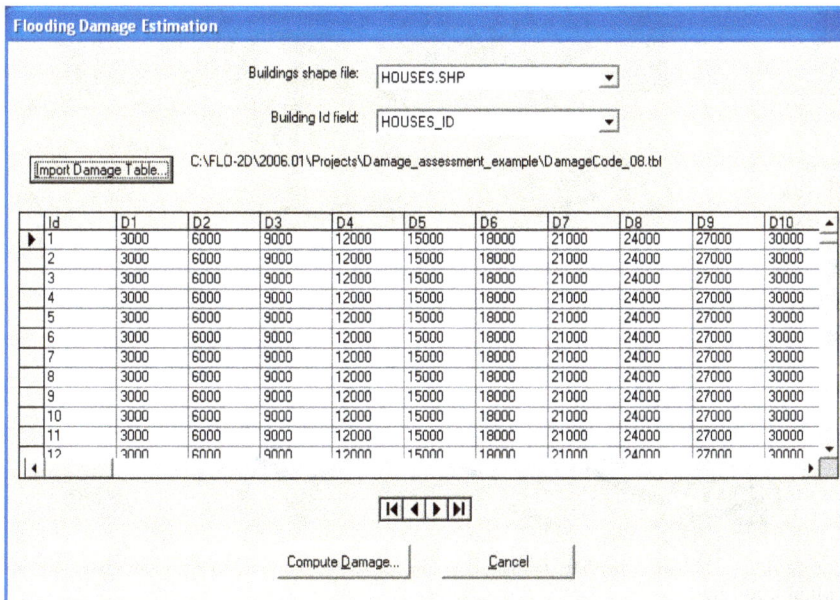

Figure 5.4   Mapper damage assessment table showing cost per foot of depth for each building shape file.

FLO-2D Mapper

Total damage cost: 141,779

OK

Figure 5.5   FLO-2D Mapper computed total damage cost for the simulated flood event.

on the up fan side. The structure details represented in the damage tables will increase the reliability of the risk mapping.

The final product for this alluvial fan flood hazard and risk analysis is a suite of maps. The first set of maps is the flow hydraulics for existing

Figure 5.6   FLO-2D Mapper displayed color coded assignment of damage cost to individual buildings.

Figure 5.7   Interpolated damage inundation cost for individual structures computed by FLO-2D Mapper.

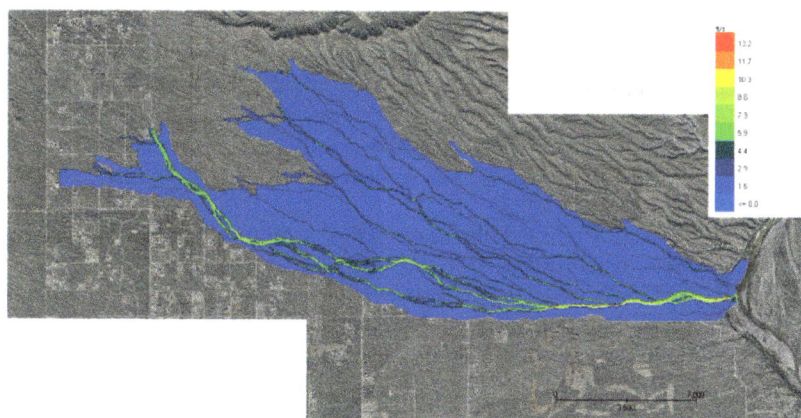

Figure 5.8   FLO-2D predicted 100-year flood maximum depths for Rio Verde alluvial fan, Phoenix, Arizona (from Regional Flood Control District of Maricopa County).

conditions (Figure 5.8) and potential future condition scenarios. This set of maps can include:

(1) Maximum flow depth (area of inundation);
(2) Maximum flow velocity;
(3) Maximum impact and static pressure;
(4) Maximum specific energy;
(5) Time to peak;
(6) Time to 1 foot; and,
(7) Time to 2 foot.

The second map is the flood hazard map based on flood intensity and flood frequency (Figure 5.1). The third map is the color coded damage costs and the associated individual structure costs (Figure 5.6). These three sets of maps and supporting modeling results provide the floodplain manager with all the tools necessary to regulate alluvial fan development through zoning, mitigation or accurate flood insurance rates based on actual damage/cost risk.

References

California Dept. of Water Resources. (2008). "Central valley floodplain evaluation and delineation project scoping document and strategic approach to hazard identification." Sacramento, CA.

Calvo, B. and Savi, F. (2008). "A real-world application of Monte Carlo procedure for debris flow risk assessment." Computers and Geoscience, Elsevier Science (in press).

Demeritt, D., Cloke, H., Pappenberger, F., Thielen, J., Bartholmes, J., Ramos, M. (2007). "Ensemble predictions and perceptions of risk, uncertainty, and error in flood forecasting." *Environmental Hazards, Elsevier Press*, 7, 115–127.

French, R.H. (1992). "Preferred directions of flow on alluvial fans." *ASCE Journal of Hydraulic Engineering*, 118, 1002–1013.

Gamma, P. (2000). dfwalk — Ein Murgang-simulations programm zur Gefahrenmoniorung. Gcographica Dernensla G00, 144 S.; Bern. — in German.

Heggan, R.J. (1994). "Determination and use of avulsion probabilities on alluvial fans." Draft Interim Procedure, Dept. of Civil Engineering, Univ. of New Mexico.

IUGS (International Union of Geological Sciences) (1997). "Working group on landslides, committee on risk assessment, quantitative risk assessment for slopes and landslides: state of the art." Proceedings International Workshop on Landslides Risk Assessment, Ed. by Cruduen, D.M., Fell, R., Rotterdam, Netherlands, pp. 3–12.

Jenson, S.K. and Domingue, J.O. (1988). Extracting topographic structure from digital elevation data for geographic information system analysis. In: Photogrammetric Engineering and Remote Sensing, Bd. 54, Nr. 11: 1593–1600.

Lind, N., Pandey, M. and Nathwani, J. (2008). "Assessing and Affording the Control of Flood Risk." Structural Safety, Elsevier Science (in press).

Maatay, J. and Maroko, A. (2008). "Mapping urban risk: Flood hazards, race & environmental justice in New York." Applied Geography, Elsevier Science (in press).

Plate, E.J. (2002). "Flood risk and flood management." *Journal of Hydrology*, 267, 2–11, Elsevier Science, B.V.

Price, W.E. (1974). "Simulation of alluvial fan deposition by a random walk model." *Water Resources Research, American Geophysical Union*, 10(2), 263–274.

Romanowicz, R. and Beven, K. (2003). "Estimation of flood inundation probabilities as conditioned on event inundation maps." *Water Resources Research*, 39(3), 1073–1085.

Scawthorn, C., Flores, P., Blais, N., Seligson, H., Tate, E., Chang, S., Mifflin, E., Thomas, W., Murphy, J., Jones, C. and Lawrence, M. (2006). "HAZUS-MH Flood loss estimation methodology. II: Damage and loss assessment." *Nat. Hazards Rev.*, 72–81.

UN-ISDR. (2004). "Living with risk. A global review of disaster reduction initiatives." Inter-Agency Secretariat of the International Strategy for Disaster Reduction. http://www.unisdr.org/eng/about isdr/bd-lwr-2004-eng.htm.

van Mamen, S.E. and Brinkhuis, M. (2005). "Quantitative flood risk assessment for polders, engineering reliability and system safety." *Elsevier Press*, 90, 229–237.

Wichmann, V. and Becht, M. (2003). "Modeling of geomorphologic processes in an alpine catchment." Conf. Proc. GeoComputation 2003, Southampton, UK, http://www.geocomputation.org/2003/index.html.

Zhao, B. and Mays, L. (1996). "Uncertainty and risk analyses for FEMA Alluvial-fan method." *ASCE Journal of Hydraulics*, 122, 325–322.

Chapter 6

Playa Lake Hazards and Resources

Julianne J. Miller

Desert Research Institute, Division of Hydrologic Sciences
755 East Flamingo Road, Las Vegas, Nevada 89119
Julie.Miller@dri.edu

Playas are common in semi- and arid environments; and their characteristic level, hard, and usually dry surfaces make them ideal for airports
and recreational uses. However, when water is present on the lakebed,
playa lakes provide habitat and a water supply for migratory birds and
nomadic herders, and opportunities for water harvest. Whereas the flooding of playas is episodic, flooding of these features can have a long duration, as typically the lakebed is impermeable and water loss occurs only
by evaporation. When water is present on the lakebed and migratory
birds are present, aircraft operations are hindered. Geologic hazards
also occur on the lakebed surfaces, including earth fissures, desiccation
cracks, and macropolygons, further restricting airfield and other lakebed
operations.

6.1 Introduction

Playas, or when wet, 'playa lakes' (terminal lakes), are common within
topographically closed basins in most semi- and arid-environments [Goudie,
1991]. They are unique hydrologic and geologic features, as both
groundwater discharge and surface water runoff can accumulate on the
lakebed surfaces. Because of this, playas are the terminus for all suspended
particles and dissolved constituents carried by water to the lakebed. Playas
often flood and estimation of inundation depth, duration, and frequency
are difficult because the watersheds within the closed basins are very large,
geologically complex, with varied land use; precipitation is spatially variable; a small amount of water can spread over the large flat lakebed; and

the strong, consistent winds that occur across playa surfaces can rapidly move water over the lakebed surface. Therefore, playas are excellent study area for geomorphology, hydrology, hydrogeology, sedimentation, and environmental toxicology [Easterbrook, 1969; Flint and Skinner, 1977].

The terms 'playa' or 'playa lake' are often confused or inappropriately used, or have been replaced by a somewhat synonymous term in a specific locale, such as 'dry lake' that is used in California [Briere, 2000; Rosen 1994]. In other parts of the world, other names refer to the same features, such as 'pans' in South Africa, 'Qa's' in Jordan, 'kahabra' in Saudi Arabia, and 'sabkha' in North Africa and portions of the Middle East.

This chapter discusses methods to determine playa lake inundation depth, duration, and frequency; geologic hazards that form on the lakebed surface; and water resource management issues associated with playas. Playa lake inundation can be determined using an innovative method used to estimate regulatory flood hazards on Rogers Lake at Edwards Air Force Base (EAFB), California, using a watershed scale model based on variations of precipitation and temperature with elevation, soil types, vegetation types and densities, and land use. In addition, remotely sensed data provided by the US Army Corps of Engineers (USACE) Cold Regions Research and Engineering Laboratory (CRREL) and precipitation data are presented to examine the frequency and duration of inundation of water on Rosamond Lake using remotely sensed data in the infrared bands.

Earth fissures, desiccation cracks, and macropolygons are common features on playa lakebeds. However, when fissures transect lakebed runways, they become hazards to aircraft movement and restrict other lakebed operations, as has occurred at EAFB beginning in the spring and summer of 1963. Three main causes of these features have been identified as: (1) groundwater lowering, (2) tectonic activity, and (3) long-term drought.

Another aspect of playas is their importance in water management on a worldwide basis. On the playas (Qa's) of the northeastern Badia region of Jordan, the issue is not primarily flooding but understanding the hydrology to provide for sustainable development while maintaining the traditional herding lifestyle of the Bedouins.

6.1.1   *Historic role of playas in military and civilian use*

Playas have played important roles in human activities in arid lands, and especially military activities, for centuries [Neal, 1968]. General Patton trained his World War II armored units on Palen Lake in the Mojave Desert

(California). Playas also have been used for military airfield and ground operations throughout the world. At EAFB, the two largest playas — Rogers and Rosamond — are currently, and historically have been (since approximately 1942), used as runways, taxiways, and industrial areas by the US Air Force (USAF) and the National Aeronautics and Space Administration (NASA) for both air and spacecraft operations (Figure 6.1). Flooding of the playas on EAFB can adversely impact flightline operations, as it can

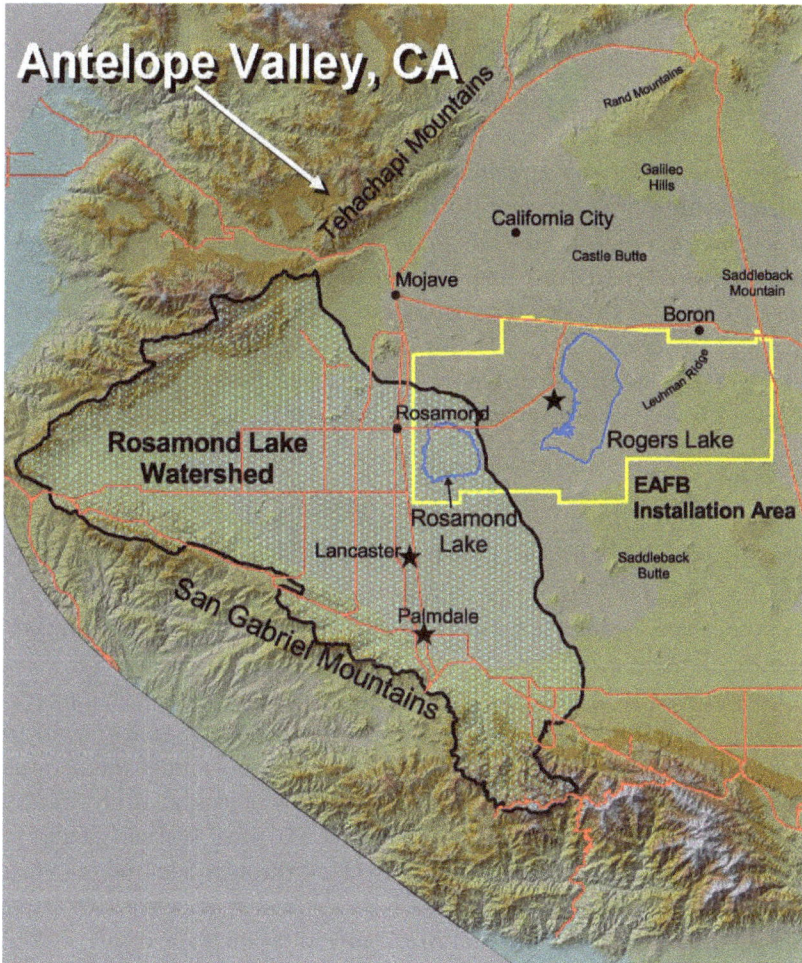

Figure 6.1   Rogers Lake and Rosamond Lake are the two largest playas on Edwards Air Force Base (EAFB), California.

at any other airport located on a playa lake. For example, runways, taxiways, aprons, and buildings are either on or only slightly elevated above Rogers Lakebed; and if the depth of inundation exceeds the elevation of these facilities then operations must cease and infrastructure damage may be incurred. Ponded water also attracts migratory birds that are a danger to aircraft operations and vice versa.

6.2   Inundation of Playas

Playa surface conditions are dictated by whether inundation occurs as surface water or by groundwater upwelling. Playas with hard surface crusts typically are found in areas where groundwater depths are well below the surface, precluding any upwelling or capillary rise of water to the surface. Surface water movement over the playa deposits creates the flat surface of the lakebed. Evaporite minerals build up on the surfaces of playas where groundwater discharge or capillary rise occurs [Neal, 1968], typically creating salt crusts. If the salt crust is regularly flooded by surface water, dissolving the minerals into a smooth surface, a salt flat, such as Bonneville Salt Flats, Utah, is created; however, if not, a rugged crust is formed, such as that of Devil's Golf Course, in Death Valley, California.

6.2.1   *Predicting the depth of inundation on playa lakes*

A flood assessment of Rogers Lake at EAFB was mandated by USAF and federal regulations, requiring the delineation of 100-year flood hazard zones. However, there is neither guidance nor a generally accepted approach to quantifying flood hazards on playas.

Edwards AFB is located in the Mojave Desert, which is characterized by long, hot summers and short, cool winters. The mean annual estimated temperature at EAFB is 16.8°C (62.2°F), although summer temperatures often reach and exceed 37.8°C (100°F). The average annual precipitation is 132.8 mm (5.23 in), measured at the EAFB precipitation gaging station located near Rogers Lake. Summer (May through September) precipitation in the Mojave Desert generally is the result of convective storms, whereas winter (October through April) precipitation is the result of frontal storms [Miller and French, 2002]. Summer convective precipitation events are intense and variable in both space and time. Winter frontal precipitation events are less intense, longer in duration, and are, generally, regional.

Winter period precipitation results in the largest rainfall amounts [Miller and French, 2002]. Therefore, the playas are more likely to be flooded by winter period precipitation events [Dinehart and McPherson, 1998; French *et al.*, 2003; 2004].

A method, described in the following text, was developed [Miller, 1998; French *et al.*, 2006] to predict the depth of playa lake inundation. Although this section focuses on use of the method to estimate the 100-year inundation depth on Rogers Lake, it can also be applied to estimate depths of events with return periods of other than 100-years. The steps in the approach and the results of each step for the EAFB analysis are summarized:

(1) Typical of semi- and arid environments, the 100-year regulatory flood event is estimated using the precipitation event with a 100-year return period. This involves the tacit assumption that there is a direct correspondence between the return periods of precipitation and runoff events. At EAFB, the regulatory 100-year precipitation event is 90.2 mm (3.55 in) in 24 hours [Miller and French, 2002].

(2) The tributary watersheds must be delineated, which often is not easy given the large watersheds associated with playas, and the complex geology of the basins.

(3) In the mountainous western United States, as is the case elsewhere, precipitation varies with elevation [French, 1983; Osborn, 1984], with the depth of precipitation generally increasing with elevation. Therefore, a location-specific relationship between the depth of precipitation and elevation for the causative precipitation event must be developed. A similar relationship between the annual average temperature and elevation is also needed.

(4) As elevation increases, the relationship between rainfall and runoff also changes because of changes in soil types and thicknesses, land use, and vegetative types and densities. The Natural Resources Conservation Service, formerly the Soil Conservation Service (SCS), curve number (CN) approach to estimating the initial abstraction of precipitation, infiltration, and runoff was used in this analysis [USDA, 1984; 1986]. The watershed tributary to Rogers Lake was divided into subbasins based on elevation, vegetative type, and vegetative cover. For each elevation interval, the threshold precipitation (depth of precipitation that must occur before any runoff takes place) was estimated based on the causative event depth of precipitation and the CN.

(5) The goal of the playa lake flood hazard analysis model is to estimate the total volume of water delivered to the lakebed given the design precipitation event; then, if the lakebed topography is known, the water depth can be estimated. Conceptually, if the elevation intervals derived in Step 4 are viewed as a series of leaky buckets, one above the other, which tip once the threshold precipitation is exceeded, the water cascades from the top of the watershed to the lakebed. The buckets leak because of the initial abstraction and infiltration, accounted for with the CN, and channel transmission losses, taken into account using the SCS climate index approach [Mockus, 1972].

Additional information must be considered in estimating the regulatory 100-year depth of water on a playa. Although playas are essentially flat surfaces with minimal topographic relief, there are always undulations and depressions present [Dinehart and McPherson, 1998]. Therefore, the accuracy of the estimated 100-year flood depth depends on the resolution of the topographic data on both the lakebed and in the immediate surrounding area. Also, playas are typically found in areas subject to strong, consistent winds that can rapidly move water over the lakebed. Therefore, two inundation depth estimates should be considered: a flat pool elevation and a location specific estimate that includes the wind set up. Although wind set up is a recognized problem in inland lakes and reservoirs [Bretschneider, 1966; Linsley and Franzini, 1979; Saville *et al.*, 1962], there have been no quantitative studies on playa lakes. Although this effect was considered in delineating the regulatory floodplain at EAFB [French *et al.*, 2003], it is not discussed in this chapter because the wind effects on the regulatory depth of water were not significant.

6.2.2 *Predicting the duration of inundation on playa lakes*

The duration of water on the lakebed is a critical issue for both aircraft operations and habitat value for migrating birds using the resource. Remotely sensed data for Rosamond Lake at EAFB (Figure 6.1) provided to the authors by the USACE CRREL were used in a study to examine the relationship between winter precipitation events and lakebed inundation [French *et al.*, 2005].

The Landsat 4 and 5 Thematic Mapper (TM) format images provided by CRREL were analyzed to correlate precipitation data with the inundation of Rosamond Lake. Other investigators [e.g., Bryant and Rainey,

2002] have used other types of remotely sensed images, such as Advanced Very High Resolution Radiometer (AVHRR), to examine playa lake inundation. Although AVHRR images have a much higher temporal resolution than Landsat, the higher spatial resolution of the Landsat pixels [approximately $900\,\text{m}^2$ $(9{,}688\,\text{ft}^2)$] relative to the resolution of the AVHRR images [$<1\,\text{km}^2$ $(<0.39\,\text{mi}^2)$] make them appropriate for examining inundation on relatively small playas, such as Rosamond Lake. Bryant and Rainey [2002] noted that in using AVHRR to estimate areas of inundation on small playas often results in relatively large errors that are likely due to the difficulty in handling the lakebed edge pixels.

Remote sensing relies on measurements of the electromagnetic spectrum to characterize the vegetation indices, soil properties, surface temperatures, and hydrologic conditions. Once data have been collected from a remote sensing system, the user must interpret them for a specific application. The software used for the remote sensing analysis was *Visual_Data*, a Windows-based program that is capable of analyzing many different types of satellite imagery [Carr, 2002].

Remote sensing of water from satellites or aircraft is an exceptional method to determine playa inundation as water absorbs or reflects most wavelengths of electromagnetic energy. Only visible wavelengths (0.4 to $0.7\,\mu\text{m}$) penetrate water, and the depth of penetration is influenced by the turbidity of the water. Energy at Infrared (IR) wavelengths (0.7 to $2.35\,\mu\text{m}$) is almost entirely absorbed by the water.

The relationship between wavelength and water penetration is also seen by comparing the individual Landsat TM bands (Figure 6.2). In Band 1, Rosamond Lake has various tones of white/gray. However, the image generated using Band 1 does not distinguish surface water or soil moisture very well. In Band 7, water is distinguished by black spectral signatures; whereas moist soils are a dark gray, and dry soils are nearly white. Figure 6.2 reveals the differences in the ability of the visible band to distinguish surface water when compared to the IR image. This portrayal demonstrates that the visible energy penetrates the surface water on Rosamond and absorbs the Near Infrared (NIR) and Short Wave Infrared (SWIR) wavelengths.

The IR spectral bands are best suited to distinguish surface water, soil moisture content, and shorelines. Shorelines are difficult to define in the visible bands because those wavelengths penetrate the shallow water and are reflected from the soils on the bottom to produce signatures similar to those of adjacent soils. However, shorelines are sharply defined in the

Figure 6.2   Visual demonstration of the spectral responses to surface water between the visible and IR wavelengths.

IR bands (4, 5, and 7) because these wavelengths are absorbed by water and are reflected by the soil. Therefore, for distinguishing surface water on Rosamond Lake, the NIR and SWIR bands were chosen.

Color composite ratio images were produced by combining three ratio images in blue, green, and red. In the color composite images for Rosamond Lake, blue was assigned Band 4 (0.76–0.90 μm), green was assigned Band 5 (1.55–1.75 μm), and red was assigned Band 7 (2.08–2.35 μm). The newly produced image (Figure 6.3) demonstrates that accumulated surface water colors range from royal blue for the deeper water to a light blue for shallower water. The soils that have high water content appear as a dark gray/brown color and dry soils appear to be light brown to white in color. Vegetation appears as greenish browns.

This study relied on the temporal resolution of the available imagery to determine the correlation between changes in surface water area observed in the images with precipitation data. Image subtraction followed by binary

Figure 6.3 Visual demonstration of the color composite generated from using Band 4 (Blue), Band 5 (Green), and Band 7 (Red) for January 8, 1985. Blue spectral responses indicate water, dark gray spectral responses indicate moist soils, white spectral responses indicate dry soils, and brown/green spectral responses indicate vegetation.

thresholding were the remote sensing processing methods used to extract the exact amount of change in surface water in the images of Rosamond Lake. Image subtraction involved subtracting one image from another to determine the difference in surface water area. Using the mosaics, the image subtraction could be checked to accurately visualize the hydrologic changes and then be further analyzed. The following subtraction order was always used, (image at time t) minus (image at time t-1).

Following image subtraction, a binary threshold was applied to delineate change in surface water area. A binary threshold is a form of contrast enhancement that changes pixel values to either zero (black) or 255 (white) depending on whether original pixel values are below (zero) or above (255) a specified threshold. For example, using the visualization tool within *Visual_Data*, the subtracted image (Figure 6.4) for the period from

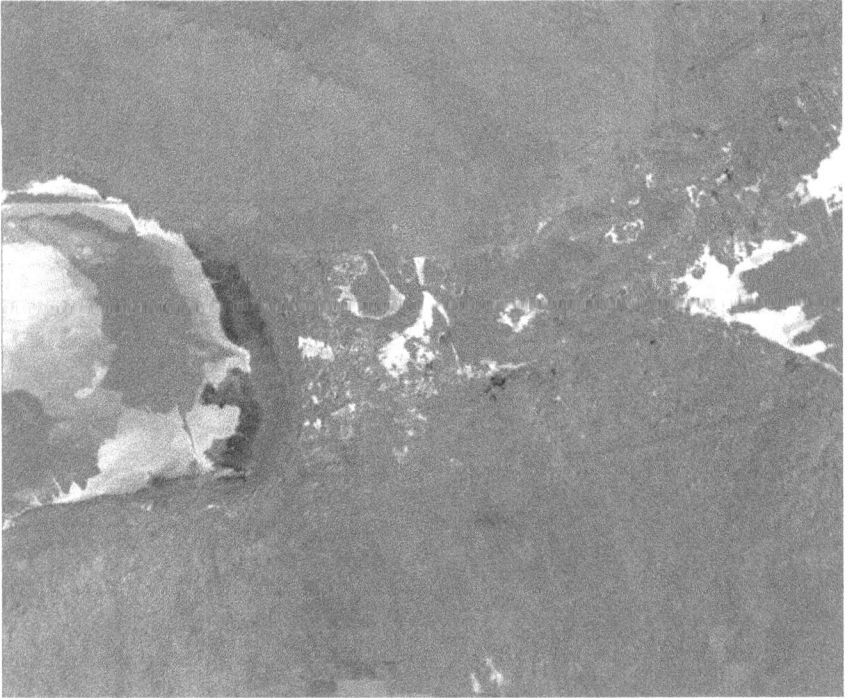

Figure 6.4   Visual demonstration revealing the image after the image subtraction was applied for the temporal period of December 13, 1992–February 15, 1993. The white to light gray areas correspond to the areal surface water changes.

December 13, 1992, through February 15, 1993, revealed that the threshold pixel value for differentiating the areal extent of water was 190. Therefore, all pixel values equal to or greater than 190 were set to a value of 255. The total number of 255 pixel values was obtained from the histogram and taken to represent the total change in surface water area between images.

The areal changes in surface water over a period of time were the focus of this study. The changes in surface water area are determined by the difference in pixels that have the spectral signature of water in the images. An image consists of the number of rows by the number of pixels per row. A pixel is one number (digit) making up a digital image. In the case of Landsat TM images, each pixel value represents the relative strength of the reflection over a 30 m × 30 m (98 ft × 98 ft) spatial region [120 m × 120 m (394 ft × 394 ft) in the case of Band 6], which is the spatial resolution of the Landsat TM satellite. Moreover, each pixel is integer-valued, represented by a single byte (8 bits), with Landsat TM pixels ranging in value from 0 to 255 (for a total of 256 possible values) [Carr, 2002].

The changes in the water surface area on Rosamond Lake were compared to the local precipitation data to determine if remote sensing can be used to establish the relationship between precipitation and the volume of water in Rosamond Lake. The previously discussed method to determine the volume of water on the lakebed was used in this study to determine the threshold precipitation depth [16.0 mm (0.63 in)] for the tributary watershed. During a precipitation event (defined as one or more consecutive days on which precipitation occurred), if the threshold precipitation is not exceeded, runoff from the watershed would not occur, but precipitation over the lake itself must be considered.

The results of the image subtraction analyses are shown Figure 6.5, with the average depth of precipitation for the time period between the remotely sensed images plotted as a function of the change in the Rosamond Lake water surface area. In general, this figure shows a relationship between the average depth of precipitation and the change in water surface area.

As expected, when the precipitation is minimal, the water surface area decreases, and when the depth of precipitation exceeds the threshold depth, the water surface area increases. The following examples are just two of the 17 image analyses completed during this study.

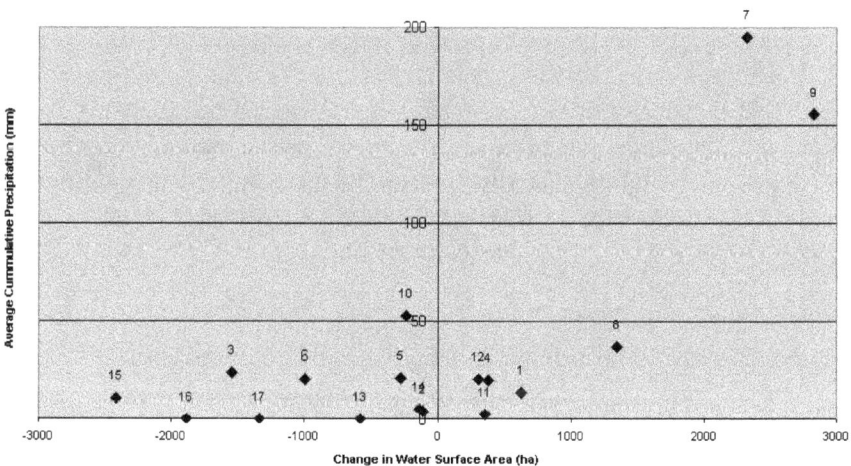

Figure 6.5 Graph demonstrating the association between surface water areal changes and averaged cumulative precipitation over each temporal period.

Figure 6.6   Visual demonstration of the color composite mosaic image for the period from January 8, 1985–February 25, 1985. The average precipitation over the period was 5.3 mm (0.21 in). This image was used to verify the accuracy of the binary thresholding procedure.

January 8–February 25, 1985

Figure 6.6 shows that the image recorded on January 8, 1985, had more area covered with water than the image recorded on February 25, 1985, as a net decline occurred in the surface water on Rosamond Lake. The southeast section of Rosamond Lake appears to be a dark gray with some standing water on January 8, 1985; whereas, the February 25, 1985, image shows an increase in dry soils in the southeast section and the western section of the lake. The subtraction between these images showed a decrease of 133.5 hectares (0.5 mi^2) in the areal extent of water on Rosamond Lake.

During the period between these images, the regional average precipitation (EAFB, Lancaster, and Palmdale) depth was 5.3 mm (0.21 in). None of the precipitation events occurring during the period exceeded the threshold depth of precipitation, so runoff was not expected; therefore, a decrease in the amount of water on Rosamond Lake is a reasonable result as both evaporation and infiltration were taking place.

March 19–April 4, 1993

Figure 6.7 demonstrates that the image recorded on March 19, 1993, had less surface water when compared to the image recorded on April 4, 1993, which indicates that the surface water area of Rosamond Lake increased. In the March 19, 1993, image, Rosamond Lake has substantial surface water inundation with dark gray (moist soil) areas mixed with surface water in the southeastern section of the lake. The April 4 image shows an increase

Figure 6.7 Visual demonstration of the color composite mosaic image for the period from March 19, 1993–April 4, 1993. The averaged precipitation over the period was 20.6 mm (0.81 in). This image was also used to verify the accuracy of the binary thresholding procedure.

in surface water in the southeastern section of the image with a decrease in the dark gray soils. The subtraction between these images showed an increase of 305.1 hectares (1.2 mi^2) in the areal extent of water on Rosamond Lake. Although the increased areal change is minimal, the area of surface water is more sharply defined in the April 4 image.

During the period between these images, the regional average precipitation depth was 20.5 mm (0.81 in). The threshold depth of precipitation was exceeded at both the EAFB and Lancaster stations during approximately the same four-day period in March; therefore, some portion of the tributary subbasin likely contributed water to the lake.

Given the foregoing, it would be expected that the water surface area on Rosamond Lake would only slightly increase, as the precipitation events that exceeded threshold depth occurred approximately one week after the first image was taken, and during a time period when evaporation rates were increasing, and infiltration was occurring, between image dates.

In the EAFB studies described here, a method to estimate the flow of water to Rosamond Lake using the SCS CN approach was hypothesized. Landsat remotely sensed multispectral images provided by CRREL and precipitation data were used to investigate the ability of the proposed method to predict the flow of water to Rosamond Lake. This study did not consider channel transmission losses, which can be significant; the effect of wind on the location and depth of water on the lakebed; evaporation of water from the lakebed, which can be significant in the spring and summer months; or infiltration of water on the lakebed, which is likely relatively insignificant. The study was also limited by the remotely sensed images provided by CRREL, as the images had been cropped so that a small portion of the lake was not visible and the elapsed time between images was often a month or more.

The results of this study demonstrate that with refinement the hypothesized method can be used to accurately predict playa lake inundation and the duration of that inundation. In general, precipitation events that should cause water to flow to the lake did; and precipitation events that should not cause flow did not. If the precipitation events that cause flow to playas can be identified, then using the same method, but incorporating channel transmission losses, the volume of flow can be estimated [Miller *et al.*, 2002]. The timing of flow and the volume can then be used with estimated evaporation rates to estimate the time water will be present on the lakebed. Therefore, this approach could prove to be a valuable tool for estimating the frequency and duration of playa lake flooding, assisting in the planning and assessment of proposed lakebed activities and facilities, and in facilitating wildlife management issues.

6.3   Geologic Hazards on Playa Lakebeds

Earth fissures, desiccation cracks, and macropolygons are common features on playa lakebeds. The US military has conducted extensive aerial and ground truthing reconnaissance of playa surfaces on which to establish airfields [Neal, 1968; Motts, 1970]. During these surveys, Neal and Motts [1967] noted an increase in the number of giant desiccation polygons ('macropolygons') occurring on lakebeds in the desert southwest US. They attributed the desiccation features to both natural climatic changes (long-term drought cycles), as well as to anthropogenic activities such as groundwater pumping, which reduces soil moisture and compacts soils, resulting in land subsidence. Pumping of groundwater for agricultural and industrial uses near the south end of Rogers Lake, EAFB, is thought to be the cause of the continuing fissuring across that lakebed, resulting in a large field of macropolygons [Blodgett and Williams, 1992; Dinehart and McPherson, 1998]. However, tectonic activity, common to the Basin and Range Province of the US, can also result in earth fissures occurring across playa lakebeds. The October 1999 Hector Mine earthquake resulted in fissuring of 150 ft (48 m) on Lavic Lake, California [Morrision, 2002]. Similarly, Yucca Lake, Nevada, was ruptured by tectonic activity in 1963, with the resulting 1 km

(0.6 mi) long fissure draining the lakebed overnight [Neal, 1968; Doty and Rush, 1985]. Many of these fissures are 1 to 10 m (3 to 32 ft) deep, and range up to 1 km (0.6 mi) in length.

6.3.1   *Evolution of desiccation cracks on playas*

Desiccation cracks develop as subsurface sediments dry and shrink, resulting in subsurface voids (Figure 6.8). When the lakebed is saturated or ponded, surface rupture is possible as the sediments become more plastic in nature, allowing the surface to collapse into the void [Neal, 1968; Messina *et al.*, 2005]. New fissures are often observed after precipitation events. Macropolygons form when desiccation cracks intersect at 90° to

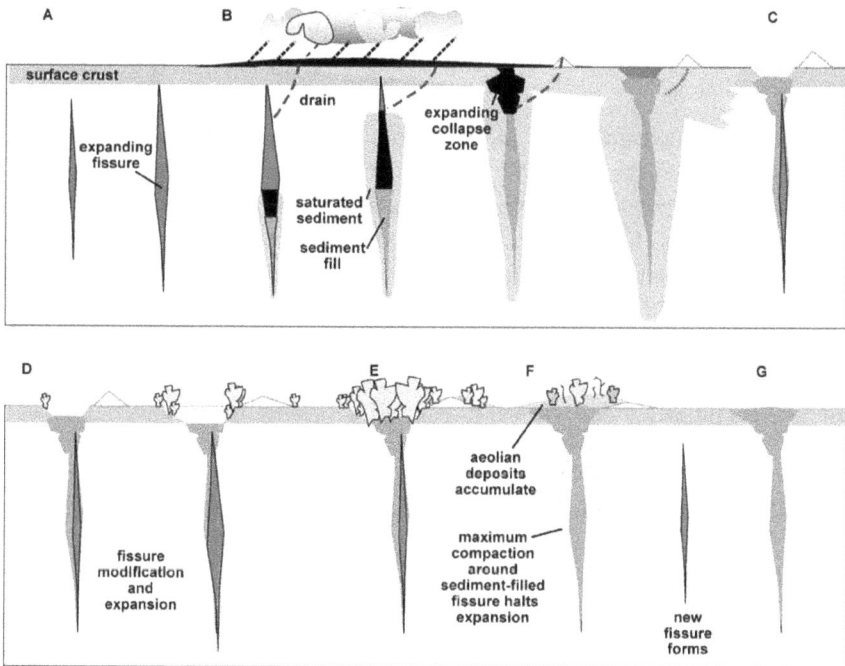

Figure 6.8   Evolution of fissure development on a playa lakebed. (A) Desiccation causes subsurface sediments to contract. (B) Ponding on the playa lakebed can cause surface rupture and inflow of water to the void. (C) The fissure drains. (D) Vegetation establishes along the margins of the fissure, which collects sediments and water. (E) Vegetation collects aeolian deposition. (F) Fissure begins to "heal" as sediment fills the void. (G) The surface is smoothed by wind and water. Modified from Messina *et al.* [2005].

Figure 6.9   A young fissure (left) and a healed fissure (right) at the south end of North Panamint Playa, California. From Messina *et al.* [2005].

120° angles, and can range in widths up to 300 m (1000 ft) [Neal *et al.*, 1968].

A natural evolution occurs as the newly formed fissures capture sediments and water (Figure 6.8). Plants become established along the margins of the fissures relatively soon after formation, and accumulate aeolian deposition, which later acts to seal or "heal" the fissure. As the amount of captured water declines in the healing fissure, plant growth begins to decrease. Eventually the scar is healed as plants die, erosion fills the fissure, and the surface is smoothed over by wind and water (Figure 6.9).

6.4   Playas as a Water Resource: Studies in Jordan

The third section of this chapter discusses how playas (Qa's) influence the surface hydrology and water management in the heterogeneous landscape of the northeastern desert of Jordan, known as the Badia Region of Jordan, which is of significant hydrologic and ecologic importance to Jordan. In the arid Middle East, playas have been utilized by both modern Bedouins and ancient people for a water supply of opportunity and convenience.

Because of population growth in recent years and the resultant increase in demands on natural resources, including water, the Jordanian government has focused on the development of the Badia Region (Figure 6.10),

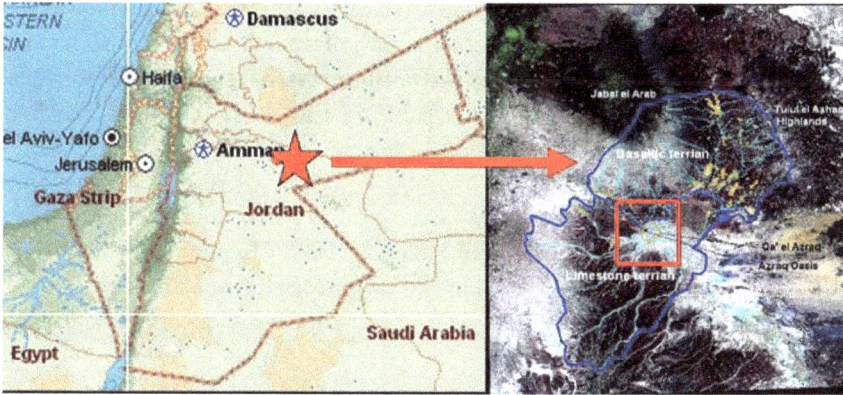

Figure 6.10 Satellite image of the Azraq Watershed with the studied playas (Qa's) delineated. These playas are primarily in the basaltic terrain.

establishing the Badia Research and Development Program (BDRP) in 1992. The objectives of the BDRP include providing for sustainable development in the desertified Badia environment and improving the standard of living for its inhabitants [Al-Homoud *et al.*, 1995]. One of the primary objectives of the BDRP is to promote scientific research of the physical environment of the Badia and build a database containing hydrologic, geologic, geomorphologic, economic, and social data for the region.

Two dominant landscape forms characterize the arid, basaltic plateau in the northeastern Badia Region of Jordan (Figure 6.10). The first is the basaltic stone pavements that cover the majority of the ground surface in the volcanic landscape. The second is the drainage network patterns, which are punctuated by sedimentary pans (playas or Qa's) of different sizes and shapes. The Qa's in the northeastern Badia Region of Jordan are the focus of this section of the chapter because of their relatively large areal extent and known significant impact on the surface water hydrology and geomorphology of the landscape. The main goal of this section is to emphasize how Qa's influence the surface water and water management in the heterogeneous landscape in the northeastern desert of Jordan, specifically in the Azraq basin.

6.4.1 *Azraq basin*

The historical importance of the Azraq watershed dates before Nabetean times (1,000 B.C.), when settlement previously had been established. The

Azraq basin includes an oasis, where the castle, Qasr Azraq, was built in the third century A.D., and was later substantially modified by the Mameluks in the Middle Ages. Qasr Azraq was an important headquarters for T.E. Lawrence (*aka* Lawrence of Arabia) during the Arab Revolt in the first quarter of the Twentieth century.

The Azraq basin is an approximately 12,177 km^2 (4,702 mi^2) topographically closed basin that lies predominately in central and eastern Jordan (Figure 6.10). The basin is within the arid Badia Region of Jordan, and receives an average of 90 mm (3.60 in) of rain per year. The relief of the basin is approximately 1,500 m (4,920 ft), with the highest elevation being the top of Jabal al Arab in southern Syria, and the lowest point being at Qa' el Azraq, at an elevation of 550 m (1,804 ft) above mean sea level.

The geology of the basin consists of two basic lithologies. To the north of the Qa' el Azraq, Tertiary to Quaternary basaltic lava flows extend all the way into southern Syria. To the south, Late Cretaceous and Paleocene limestone and chert beds from the Um Rijam formation, and other recent limestone formations, are predominant. The diverse geology influences the geomorphology and hydrology of the basin.

The geomorphology and surface hydrologic characteristics of the basalt flow surfaces play a critically important role in the hydrologic balance of the area. Al-Qudah [2003] demonstrated that although some recharge occurs through the basaltic stone pavements in the higher elevations of Tulul al Ashaquif, most precipitation travels as surficial runoff flowing to the wadis (ephemeral channels) emanating from the area and subsequently conveying flows to the playas.

In the northern half of the basin, where basaltic lava flows cover the surface, the drainage network development varies. Although the drainage network is well-developed on older basaltic flow surfaces, it is poorly developed on most recent younger flow surfaces. Also, channel paths are directed along the edges of basaltic lava flows. Playas (Qa's) of different sizes and shapes are highly concentrated on the basaltic flow surfaces. Qa' surfaces are smooth, flat, light-colored, and mostly vegetation free. Because of the roughness of the basaltic flow surfaces and presence of the large number of playas on these surfaces, the basaltic terrain is poorly drained and significant amounts of surface water are entrapped. In contrast, there are few playas in the southern half of the basin, where limestone and alluvial sediments, as well as a chert pavement, cover the surface. The drainage networks are well-developed and wadi sub-catchments well-drained, with most of the surface water flow to Qa'el Azraq.

6.4.2 *Playas in the Northeastern Badia*

All playas of the northeastern Badia Region of Jordan are ephemeral as they are fed only by storm water. Water stands in these playas for several weeks to a few months during wet years, similar to the playas at EAFB. The only exception is at Qa' el Azraq, where groundwater from the shallow basaltic aquifer formerly discharged as springs, creating the historic Azraq oasis (Figure 6.11). However, the Azraq oasis, part of Qa' el Azraq, is now mostly dry because of the groundwater over pumping, resulting in a lowering of the water table by about 15 m (49 ft) in the last three decades.

As water flows downstream from the headwaters of the northern basaltic region of the Azraq basin, additional flow may be contributed from playas formed between volcanic cinder cones at higher elevations, as seen at Tulul el Ahsaqif highlands (Figure 6.10). As the flow continues downstream through well-defined channels it may diffuse and spread onto the large smooth Qa's where water velocity significantly drops, although ponding does not occur, allowing most of the coarse sediment load to settle out of the flow, forming what is locally called "Marab". Infiltration is high on Marab surfaces, allowing relatively dense vegetative cover. Beyond the Marab, the flow discharges to semi-closed pans, or Qa's, where water ponds and fine sediments, consisting mostly of clay, silt, and fine sand, are deposited. Depending on inundation depth, Qa's may become full, allowing water to spill over and continue downstream through the channel system. This process may repeat itself, until the flow terminates in a large, closed Qa'.

Figure 6.11   Qa' el Azraq and the Azraq Oasis are important water resources for Jordan.

6.4.2.1 *Determining water volume on playas within the Azraq basin of Jordan*

The total area of the Azraq basin is approximately $12,177 \, \text{km}^2$ ($4,702 \, \text{mi}^2$), with the basaltic flow surfaces constitute approximately 46 percent of the total area [$5,709 \, \text{km}^2$ ($2,163 \, \text{mi}^2$)] and limestone surfaces constitute approximately 54 percent of the total area [$6,468 \, \text{km}^2$ ($2,539 \, \text{mi}^2$)]. In this study, only the 65 playas greater than $100 \, \text{km}^2$ ($38 \, \text{mi}^2$) were analyzed. Approximately 92 percent of these playas were within the basaltic flow surface area, covering a total surface area of $222 \, \text{km}^2$ ($84 \, \text{mi}^2$), and ranging from 0.1 to $28 \, \text{km}^2$ (0.04 to $11 \, \text{mi}^2$) each. Qa' el Azraq, the main playa of the basin, has a surface area of $78 \, \text{km}^2$ ($30 \, \text{mi}^2$), and is located within the limestone portion of the basin, although near the basalt flows (Figure 6.10).

The basalt portion of the Azraq basin receives an approximately $110 \, \text{mm}$ (4.4 in) of precipitation per year because of the orographic effect of Jabal el Arab, where precipitation increases to approximately $350 \, \text{mm}$ (14 in). Therefore, on the basaltic surface, approximately $628 \times 10^6 \, \text{m}^3$ (509,100 ac-ft) of water is expected. Using an average runoff coefficient (the ratio of runoff to rainfall) of 0.6, the volume of runoff is reduced to $376 \times 10^6 \, \text{m}^3$ (304,800 ac-ft). Half of this volume of water is retained in playas and does not contribute to groundwater recharge as it is lost to evaporation.

The volume of retained water in the playas of the Azraq basin and the northeastern Badia Region of Jordan is considered to be significant with respect to the aridity of the region. Therefore, management of this water resource is essential both to inhabitants of the area and to Jordan as a whole. For example, if this water could be banked in the stressed groundwater aquifer, or otherwise used prior to evaporation, it could provide an important contribution to sustainable development.

6.5 Conclusions

Playas are important resources that are poorly understood from the viewpoint of water resources engineering. The prediction of the frequency, depth, and duration of playa lake inundation are critical issues from many perspectives. As playas are generally level and dry, they are excellent locations for airfield and other ground operations, especially for military exercises. However, when they are inundated, aircraft operations may be impaired. Geologic hazards such as desiccation cracks, fissures, and macropolygons also may limit lakebed airfield operations. From a water supply viewpoint, water on playas could be an important resource if that water can be harvested or used to sustain lifestyles and current and future water needs. Playas are a worldwide critically important natural phenomenon and resource.

References

Al-Homoud, A.S., Allison, R.J., Sunna, B.F. and White, K. (1995). "Geology, geomorphology, hydrology, groundwater, and physical resources of the desertified Badia environment in Jordan." *Geojournal*, 37(1), 51–67.

Al-Qudah, K. (2003). "The influence of long-term landscape stability of flood hydrology and the evolution of the valley floor in the northeastern Badia of Jordan." unpublished dissertation, University of Nevada, Reno, Reno, NV.

Blodgett, J.C. and Williams, J.S. (1992). Land subsidence and problems affecting land use at Edwards Air Force Base and vicinity, California, 1990: US Geological Survey Water-Resources Investigations Report 92-4035, 25 p.

Bretschneider, C.L. (1966). "Engineering aspects of hurricane surge." in Ippen, A.T. (ed.), Estuary and Coastline Hydrodynamics. McGraw-Hill, New York, NY.

Briere, R.G. (2000). Playa, playa lake, sabkha: proposed definitions for old terms. *Journal of Arid Environments*, 45(1), 1–7.

Bryant, R.G. and Rainey, M.P. (2002). "Investigation of flood inundation on playas within the Zone of Chotts, using time-series of AVHRR." *Remote Sensing of Environment*, 82, 360–375.

Carr, J.R. (2002). "Data Visualization in the Geosciences." Prentice Hall. Upper Saddle River, NJ.

Dinehart, R.L. and McPherson, K.R. (1998). "Topography, Surface Features, and Flooding of Rogers Lake Playa, California," Water Resources Investigations Report 98-4093, U.S. Geological Survey, Sacramento, CA.

Doty, G.C. and Rush, F.E. (1985). Inflow to a crack in playa deposits of Yucca Lake, Nevada Test Site, Nye County, Nevada. US Geological Survey Water Resources Investigation Report, 84-4296, U.S. Geological Survey, Sacramento, CA.

Easterbrook, D.J. (1969). Principles of Geomorphology. McGraw-Hill Book Company, New York, NY.

Flint, R.F. and Skinner, B.J. (1977). Physical Geology, Second Edition. John Wiley & Sons, New York, NY.

French, R.H. (1983). "Precipitation in Southern Nevada." ASCE, *Journal of Hydraulic Engineering*, 109 (7), 1023–1036.

French, R.H., Miller, J.J., Dettling, C.R. and Carr, J.R. (2006). "Use of remotely sensed data to estimate the flow of water to a playa lake," *Journal of Hydrology*, 325, 67–81.

French, R.H., Miller, J.J. and Dettling, C.R. (2005). Estimating playa lake flooding: Edwards Air Force Base, California, USA. *Journal of Hydrology*, 306, 146–160.

French, R.H., Miller J.J. and Dettling, C.R. (2004). "Flood Assessment for Rosamond Lake, Edwards Air Force Base, California." Division of Hydrologic Sciences, Desert Research Institute, Las Vegas and Reno, NV.

French, R.H., Miller, J.J. and Dettling, C. (2003). "Flood Assessment for Rogers Lake, Edwards Air Force Base, California." Publication No. 41185, Division of Hydrologic Sciences, Desert Research Institute, Las Vegas and Reno, NV.

Goudie, A.S. (1991). "Pans," *Progress in Physical Geography*, 15(3), 221–237.

Linsley, R.K. and Franzini, J.B. (1979). "Water Resources Engineering." McGraw-Hill, New York, NY.

Messina, P., Stoffer, P.W. and Smith, W.C. (2005). "Macropolygon morphology, development, and classification on North Panamint and Eureka Playas, Death Valley National Park, CA." in *Fifty years of Death Valley Research*, Calzia, J.P. (ed.), *Earth-Science Reviews*, 354p.

Miller, J.J. and French, R.H. (2002). "Flood Assessment for Mojave Creek, Edwards Air Force Base, California." Publication No. 41203, Division of Hydrologic Sciences, Desert Research Institute, Las Vegas and Reno, NV.

Miller, J.J. (1998). "Yucca Flat runoff, Yucca Lake Depth, Yucca Lake flood hazard." Bechtel Nevada, prepared for the US Department of Energy, Las Vegas, NV.

Mockus, V. (1972). "Chapter 21: design hyetographs" in Minor Revisions by McKeever, V., Owen, W. and Rallison, R. (eds.), National Engineering Handbook, Section 4, Hydrology, U.S. Department of Agriculture, Soil Conservation Service, Washington, D.C.

Morrison, H. (2002). Geophysical studies of the Lavic Lake Fault, Lavic Lake Playa, California. M.S. Thesis, Department of Geology, San Jose' State University, San Jose', California.

Motts, W.S. (1970). Geology and hydrology of selected playas in Western United States. Final Scientific Report, prepared by the Geology Department, University of Massachusetts, Amherst for the Air Force–Cambridge Research Labs. 286 pp.

Neal, J.T., Langer, A.M. and Kerr, P.F. (1968). Giant Desiccation Polygons of Great Basin Playas. Geological Society of America Bulletin, Vol. 79, 69–90.

Neal, J.T. (1968). Playa surface morphology: miscellaneous investigations. Office of Aerospace Research, U.S. Air Force. AFCRL-68-0133. 150 pp.

Neal, J.T. and Motts, W.S. (1967). Recent geomorphic changes in playas of western United States. *Journal of Geology* 75(5), 511–525.

Osborn, H.B. (1984). "Estimating precipitation in mountainous regions," *Journal of Hydraulic Engineering*, 110(12), 1859–1863.

Rosen, M.R. (1994). "The importance of groundwater in playas: A review of playa classifications and the sedimentology and hydrology of playas." Paleoclimate and Basin Evolution of Playa Systems, Geological Society of America, Special Paper 289, 1–18.

Saville, T., McClendon, E.W. and Cochran, A.L. (1962). "Freeboard allowances for wave in inland reservoirs," *Journal of the Waterways and Harbors Division*, 88(WW2), 93–124.

U.S. Department of Agriculture (USDA) (1986). "Urban hydrology for small watersheds." Technical Release 55, 2nd Edition, U.S. Department of Agriculture, Soil Conservation Service, Washington, D.C.

U.S. Department of Agriculture (USDA) (1984). "Engineering field manual for conservation practices." U.S. Department of Agriculture, Soil Conservation Service, Washington, D.C.

Chapter 7

Needs and Benefits of Co-Operation

Richard H. French

Department of Civil & Environmental Engineering
University of Texas at San Antonio
6900 N Loop 1604 West, San Antonio, Texas 78249
Richard.French@utsa.edu

Jonathan E. Fuller

JE Fuller/Hydrology & Geomorphology, Inc.
8400 S Kyrene Rd., Suite 201, Tempe, Arizona 85284
jon@jefuller.com

Philip J. Shaller* and Parmeshwar L. Shrestha[†]

Exponent, Inc.
320 Goddard, Suite 200 Irvine, CA 92618
*pshaller@exponent.com; [†]pshrestha@exponent.com

The identification and mitigation of flood hazard on alluvial fans is a complex task that can benefit greatly from the involvement of professionals other than civil engineers, and in most cases the economic and engineering benefits of doing this far outweigh the costs [French and Keaton, 1992]. In this chapter, specific examples of cross-discipline and private-public cooperation are discussed.

7.1   Introduction

Projects involving flood hazard identification on alluvial fans typically begin with a data collection effort, and even at this early stage engineers can benefit from the thoughts, experience, and input of other professionals, such as geologists, geomorphologists, soil scientists, hydrologists, planners, historians, and others. Indeed, with respect to our understanding of alluvial fan flooding, the lines between traditional domains of these disciplines have blurred, resulting in valuable new insights, approaches, and understanding. Given the overlap between geologic, hydrologic, engineering and planning processes, it is no longer useful or desirable to approach alluvial fan flooding problems from the narrow perspective of a single discipline.

In the following sections, a number of examples are presented in which alluvial fan flooding investigations have benefitted from multi-disciplinary cooperation. Additional examples are provided in the case histories discussed in Chapter 8.

7.2   Identifying the Alluvial Fan Hydrologic Apex

Simply identifying the hydrologic apex of an alluvial fan requires insights and cooperation from engineers, hydrologists and geomorphologists. The hydrologic apex is defined as the point where active tributary watershed processes transition to depositional alluvial fan processes. The hydrologic apex is conceptually distinct from the geomorphic apex, which is a geometric characteristic of an alluvial fan. Although the two are frequently coincident, it is not always the case, especially in the case of fans that have been entrenched in response to either climatic or tectonic factors.

Once the hydrologic apex is defined, a hydrologist may be called on to estimate peak discharge rates, hydrograph volumes, and flow frequency curves. These hydrologic data are then used by the hydraulic engineer to estimate channel capacity and predict flow depths and velocities on the fan surface below the apex. Sedimentation engineers build on the hydrologic data and channel hydraulics to predict how sediment yield may impact channel capacity, locations of net or episodic sediment deposition, and whether scour impacts channel capacity estimates. Geomorphologists provide additional insight on whether debris flows should be considered, and interpret surficial landforms to provide evidence of past flood processes, which also may be used to verify or calibrate engineering and hydrologic data. Geologists provide insights on how fan aggradation processes and rates may affect the future location of the hydrologic apex. Planners and agency personnel provide a regulatory context in which decisions relating to acceptable levels of risk and design standards can be made. The most robust analyses include a feedback loop in which evidence from all disciplines is used to fine-tune and revise the results.

7.3   Watershed Delineation

The watershed above the hydrologic apex of the fan is usually delineated using Digital Elevation Maps (DEM's) and the algorithms of a Geographic Information System (GIS). In general, the results obtained in this fashion are satisfactory. However, in some cases the engineer must exercise judgment and oversight of automated delineation processes if the topography is complex. For example, the southwestern boundary of the Rosamond Lake, California [French *et al.*, 2006] watershed occurs at the confluence of the San Andreas Fault, a major aqueduct, and a water supply reservoir. The required delineation was accomplished by reviewing topographic maps, aerial photographs, and conducting field inspections. In this case, the watershed delineation was improved by employing experience in map interpretation, field judgment, and local knowledge. In other cases, automated watershed delineation tools may be confounded by distributary flow paths, poorly defined divides, hummocky or karst terrain, agricultural or urban development, elevated roads or railways, and low divides that are difficult to automatically identify in the absence of high definition topographic data, and are subject to overtopping at high flow rates. The lateral topographic relief below the hydrologic apex generally decreases at an exponential rate with distance from the apex. As a result, delineating the lateral fan boundaries (especially in distal reaches of a fan) can be difficult and may require a careful field inspection. For example, the Rosamond and Rogers Dry Lake, California watersheds are separated by a stable linear sand dune that is not shown on topographic maps, [French, *et al.*, 2006]. Thus, the watershed delineation was improved by considering field-based landform data and geomorphic interpretation of the dune landform stability. In addition, linear anthropogenic features such as highways, railroads, and power corridors may require consideration [Schick, 1974].

In some cases, the lateral boundaries of an individual alluvial fan, as defined by surficial age (geomorphology), may not be coincident with any measurable topographic relief. These low-relief geomorphic boundaries may or may not be meaningful when modeling future flooding potential or hydraulic processes (engineering). This situation occurs on most bajadas, especially near the toe of the piedmont. Therefore, it is necessary to explicitly identify the study objectives so that engineering and geologic information can be properly assimilated in the analysis.

7.4   History

Flood hazard identification and mitigation are performed on an engineering time scale. However, engineers should keep in mind the adage that those who forget history are fated to repeat it. Historical data sets can provide key insights to the nature of alluvial fan flood hazards within an engineering time scale. Valuable historical information can be obtained by comparing recent with older topographic maps and aerial photographs and quantifying the observed changes. Historical maps can be particularly valuable in showing landscape responses (or lack of response) to known past flood events, and might include channel avulsions, channel erosion, debris flow deposition, or channel incision, as well as human impacts such as structure damage or remedial work. There may also be valuable information and data in local publications and histories, including first-hand accounts, flood damage reports, bridge failures, or ground photographs. In some cases the data collection effort becomes an expanding treasure hunt with each piece of information leading to another valuable piece of information.

Historical information can also be used to verify or calibrate engineering models to assure that the modeling results are appropriate and reasonable. Model calibration might include hindcasting flood inundation limits with known rainfall and runoff measurements. Differences between model predictions and historical records should be fully investigated, rather than dismissed or explained away, because they may elucidate gaps in our understanding of fan flood processes, as well as test the accuracy of our predictions.

7.5   Surficial Geology

Geologic data are another form of historical records, albeit one that represents very long time spans, and can serve the same role in alluvial fan investigations as other historical data. The geologic flood histories of most fans are well preserved and can be readily observed on undeveloped fans.

Alluvial fan surfaces range in age from very young (actively flooded and aggraded) to very old (millions of years). Surface age is of interest for alluvial fan flood studies primarily because it is an indicator of stability.

That is, a geologically old surface becomes old if it has not been subject to frequent flood inundation, erosion or sediment deposition, *i.e.*, it is not in the floodplain. The fact that the surface has not been in the floodplain for thousands to millions of years, makes it unlikely that it will be flooded in the near future (aside from localized sheet flows). Note, from a purely statistical view point this inference cannot be true. Rather, the statement implies that there is a physical reason, apparent or not, why this area has not been subject to flooding. Surface age can be estimated (but rarely quantified) from field observations, aerial photographs and soils mapping. In arid climates, indicators of surface age include the presence of desert pavement, desert varnish, soil profile development, surface texture, color, vegetative cover, drainage pattern, and lateral relief [French, 1987]. These indicators provide a ready means to distinguish Pleistocene from Holocene surfaces.

Except in unusual circumstances, however, absolute age dating of Pleistocene and Holocene surfaces on arid region fans remains elusive (though new techniques are being developed that have the potential to improve this situation). Analogous indicators of surface age are also present on fan surfaces in humid and temperate climates. Here, rapid soil development and the preservation of organic material make age dating more tractable.

The bedrock and surficial geology also provides clues to the types of flood processes occurring on a fan. For example, watersheds containing exposures of fine-grained sedimentary and metamorphic rock produce large quantities of fine detritus that tends to get mobilized in mudflows and debris flows. These fans have morphologic features that are easily distinguished from fluvially-dominated fans. Furthermore, sediment sizes and bedforms provide clues about frequent or recent flow depths and velocities along channels or overbank areas. The variation of channel dimensions over the length of the fan provide information about rates of channel infiltration and flow attenuation, frequency of flow bifurcations, and impacts of tributary flow sources. Finally, new surficial dating techniques are being developed to provide age constraints that may help estimate the recurrence interval of fan changing events such as major floods, channel avulsions, or debris flows. At present, however, these techniques remain largely experimental, and are not generally amenable for standard engineering practice. Nevertheless, geomorphic interpretation of the alluvial fan surface based on field observations and proper use of aerial photographs by experienced professionals can illuminate many features on an alluvial fan directly related to flood hazard identification.

7.6 Paleohydrology

Paleohydrology can provide critical data for any flood study, and is a shared discipline between the geosciences and engineering. Although some aspects of paleohydrology will be addressed here, the reader is referred elsewhere for a more comprehensive discussion, [*e.g.*, House *et al.*, 2002]. Paleohydrologic analysis provides information about the magnitude and character of floods that pre-date modern systematic streamflow measurement. Paleohydrologic studies can be used to:

(1) Calibrate peak discharge estimates computed from hydrologic models [House *et al.*, 2002];
(2) Identify non-exceedance flow thresholds by comparing flood stage estimates, rating curves, and surficial geologic data [Stedinger and Cohn, 1986];
(3) Estimate maximum velocities using boulder transport theory [O'Connor *et al.*, 1986]; and,
(4) Identify inundation limits and flow depths from recent events.

Paleoflood investigations are somewhat more tractable on debris flow fans, where flood events can leave a record in the form of debris flow deposits. These investigations are complicated by reworking of debris and, in arid regions, by the usual complication of age dating of Holocene surfaces where the rate of soil development is slow.

An example of the opportunities and complexities of presented by the study of debris flow fans is given by Shaller *et al.* [2006], who reported on flood and debris flow recurrence at a watershed near Palm Springs, California. A key finding of this study was that while hydrologic records for the watershed suggest that storm events capable of generating debris flow activity should recur every few decades, archeological evidence suggested that the watershed had not produced a debris flow for at least 350 years. The absence of debris flows during this period suggests that their occurrence is tied to hydrologic events that are rare or absent in the current climate regime.

7.7 Aggradation and Scour

Excavating trenches on a fan surface also provides several lines of valuable information for flood hazard assessment on alluvial fans. First,

Newtonian (fluvial) and non-Newtonian (debris and mud flow) deposits can be identified by trained stratigraphers, and can help identify the types of flood hazards likely to occur in the future. Second, significant flood events in the recent geologic past can be dated and used to supplement the modern flood record. This is especially useful in remote regions where there are no modern systematic gaging records. The combination of hazard type and event frequency may be especially useful, given the likelihood that some fan processes such as avulsions and debris flows may occur on long recurrence intervals, but are so severe in consequence that they merit protection that exceeds standard risk levels. Third, the average rate of net aggradation in the active fan areas can be estimated if datable material can be identified in the soil profile. The average rate of aggradation is directly related to hazard of channel avulsion, and is important if engineered flood mitigation measures are proposed. Fourth, the maximum single event sediment deposition depth can be measured from the exposed stratigraphic column. In arid regions, where the median may depart significantly from the mean value, the maximum single event volume may be a more important design consideration than the (geologically) long-term average. Fifth, precipitation on a geologic time scale can often be estimated from subsurface information.

7.8 Climate Change

The effects of climate variability on alluvial fan flood hazards in semi and arid regions is becoming of interest, French and Irvin [2007]. For example, in the American Southwest El Nino and La Nina have long been of interest from the viewpoint of water supply [*e.g.*, Andrade and Sellers, 1988; Cayan, 1996; Cayan and Peterson, 1989; Cayan and Webb, 1992; Piechota *et al.*, 1997; and Schonher and Nicholson, 1989]. While El Nino and La Nina may have an effect on long duration winter flow events [URS, 2004], the research is in the preliminary phase. Climatic impacts on vegetative cover, wildfire frequency, seasonality of runoff, and their consequence on sediment supply may significantly alter alluvial fan flood processes over the next several decades.

7.9 Planning

Much of the planning and cost of flood mitigation structures on alluvial fans could be minimized with proper planning and in selecting a good location.

For example, the facility discussed in French and Keaton [1992] was originally sited in the bottom of an ephemeral channel 5 ft (1.5 m) deep and over 100 ft (30 m) wide, and it was obvious this was not a good location from the viewpoint of flood mitigation. The client agreed to allow the facility to be moved within a reasonable distance. The project team, which consisted of engineers and geologists, found a site a short distance away that required minimal flood mitigation construction; that is, the site was naturally protected, as demonstrated by the age of the surface and engineering calculation.

A typical and more difficult situation is designing cost-effective flood mitigation measures for existing development. This endeavor can still be aided by planning for the present and future. Expanding or improving an existing drainage system for an existing community is a challenging undertaking. In the United States, Las Vegas, Nevada is an excellent example of modifying an existing drainage system impacted by alluvial fan flooding. In Las Vegas, the problem consisted of an older, built-out metropolitan core with a drainage system that had been constructed over several decades using differing design standards. Outside the core area, new development was rapidly occurring and overloading the older core area drainage system. Two alternatives were considered to handle drainage when the outer area was completely built out in the future. The first alternative was to detain flood waters at the furthest edges of the metropolitan area, in many cases at the hydrologic apices of the surrounding fans. The second alternative was to modify the interior drainage system to convey all flood waters through the urban area. Of the two alternatives, the first was the most cost-effective since it did not involve purchasing additional right of way in the congested core area; rather, construction was confined to relatively undeveloped areas. This planning decision dictated the engineering approach used to evaluate and mitigate the alluvial fan flood hazard.

7.10 Summary

In conclusion, the goal of an engineering flood hazard identification project is the production of maps delineating flood hazards based on geologic mapping and engineering calculation. In isolated cases, this approach has resulted in maps that are not representative of what can be clearly seen in the field; that is, the maps based on engineering calculation defy common sense [Fuller, 1990]. The common purpose of engineers and geologists is to construct and present to the public the most accurate hazard map possible with the resources available. In many cases, there is a geologic hazards map which includes flood hazard and is based on geologic data and observations. It is recommended that the engineering hazard map be conservative as compared with the geologic hazard map. While differences between the maps are to be expected, there should be overall agreement in identifying high and low hazard areas. When significant discrepancies occur, it is prudent to attempt to identify the reason or reasons for the discrepancy. Finally, it must be recognized that the design life of some special engineered facilities, such high level radioactive waste storage facilities, approach a geologic time scale (10,000 years), adding substantially to the complexities of projecting evidence of past hydrologic behavior into the future.

References

Andrade, B.N. and Sellers, W.D. (1988). "El Nino and its effects on precipitation in Arizona and western New Mexico." *Journal of Climatology*, 8, 403–410.

Cayan, D.R. (1996). "Interannual climate variability and snowpack in the western United States." *Journal of Climate*, 9, 928–948.

Cayan, D.R. and Peterson, D.H. (1989). "The influence of North Pacific circulation on streamflow in the West." *Aspects of Climate Variability in the Pacific and Western Americas, Geophysical Monograph*, 55, 375–398.

Cayan, D.R. and Webb, R.H. (1992). "El Nino/Southern oscillation and streamflow in the western United States." In *El Nino, Historical and Paleoclimatic Aspects of the Southern Oscillation*, edited by H.F. Diaz and V. Markgraf, Cambridge University Press, Cambridge, England, 29–68.

French, R.H. and Irvin, J. (2007). "El Nino-La Nina Implications on Flood Hazard Mitigation — Phoenix, AZ. Area." *Proceedings of the World Water and Environmental Resources Congress*, ASCE/EWRI, Tampa Florida.

French, R.H., Miller, J.J., Dettling and Carr, J.R. (2006). "Use of remotely sensed data to estimate the flow of water to a playa lake." *Journal of Hydrology*, 325, 67–81.

French, R.H. and Keaton, J.R (1992). "Successful interactions between hydraulic engineering and geomorphology in identifying flood hazard areas in the Southwestern United States." *Proceedings of the 1992 Hydraulic Engineering Sessions at Water Form '92*, ASCE, Baltimore, MD, 581–586.

French, R.H. (1987). "Hydraulic processes on alluvial fans." Elsevier Science Publishers, Amsterdam, NL.

Fuller, J. (1990). Misapplication of the FEMA alluvial fan model: A case history. *Proceedings of the International Symposium on Hydraulics and Hydrology of Arid Lands*, ASCE National Hydraulics Conference, Boston, Massachusetts. pp. 367–372.

House, P.K., Webb, R.H., Baker, V.R. and Levish, D.R. (Eds.) (2002). "Ancient floods, modern hazards: Principles and applications of paleoflood hydrology." American Geophysical Union, Washington, D.C.

O'Connor, J.E., Webb, R.H. and Baker, V.R. (1986). "Paleohydrology of pool-and-riffle pattern, development: Boulder Creek, Utah." *Geological Society of America Bulletin*, 97, 410–420.

Piechota, T.C., Dracup, J.A. and Fovell, R.G. (1997). "Western U.S. streamflow and atmospheric circulation patterns during El Nino-Southern Oscillation." *Journal of Hydrology*, 201, 249–271.

Schick, A.P. (1974). "Alluvial fans and desert roads — a problem in applied geomorphology." *Abhandlungen der Akademie de Wissenshaften* in *Gottingen Mathematisch-Physikalische Klasse*, 29, 418–425.

Schonher, T. and Nicholson, S. (1989). "The relationship between California rainfall and ENSO events." *Journal of Climate*, 2, 1258–1269.

Shaller, P.J., Hamilton, D., Shrestha, P.L., Lyle, J.E. and Doroudian, M. (2006). "Investigation of flood and debris flow recurrence — Andreas Canyon, San Jacinto Range, Southern California." *ASCE World Environmental and Water Resources Congress, 2006*, 7 p.

Stedinger, J.R. and Cohn, T.A. (1986). Flood frequency analysis with historical and paleoflood information, *Water Resources Research*, 22(5), 273–286.

URS (2004). *Multiple Storms Analysis — On-Call Special Studies, Phase II*, Contract No. FCD 2001 C050. Prepared for: Flood Control District of Maricopa County, Phoenix, AZ.

Chapter 8

Meeting the Challenge

In this chapter, three case studies are presented which were selected because they illustrate, through application, the material discussed in the preceding seven chapters and engineering challenges requiring state-of-the-art solutions.

In Case Study #1, the assignment is to identify and mitigate flood hazard to a development in the alluvial fan environment east of San Diego, CA. This is a presentation and discussion of a typical project in which the engineer must design cost-effective drainage mitigation for the client while ensuring other properties in the vicinity are not adversely impacted and the beneficial use of land is maximized. In humid environments, this is a traditional engineering problem requiring a minimum of innovative engineering; however, as discussed in the foregoing chapters, this is not an insignificant challenge in an active alluvial fan environment where there are neither well defined nor stable channels. In this instance, the FLO-2D model was used to assess both off-site and proposed on-site drainage paths for the regulatory 100-year flow event. The use of the FLO-2D model allowed the original design to be modified to provide greater flow efficiency and convey the flow through the development at lower velocities than originally anticipated.

Case Study #2 discusses the large scale debris flow that occurred at La Conchita, CA. on January 10, 2005, tragically destroying 36 residences and killing 10 persons. The subsequent investigation and two-dimensional modeling of the event involved engineers, hydrologists, and geoscientists. The resulting model provided insight as to both the mechanics of the debris flow and its interaction with a temporary retaining wall at the site, which partially failed during the event. This section clearly indicates the necessity and importance of a truly cooperative inter-disciplinary approach to address cutting edge problems.

In Case Study #3, the Tiger Wash alluvial fan near Phoenix, Arizona, is discussed from the perspective of it essentially being a full scale laboratory experiment in progress. Data were collected at this location before and after a very large flood in 1997. The data collected allow very broad, critical, and fundamental questions to be addressed, such as what is an alluvial fan and how to identify alluvial fan boundaries. The work at Tiger Wash highlights the need for close interaction among engineers, geologists, and floodplain managers when assessing alluvial fan flood hazards.

Chapter 8 is intended to summarize the preceding material and demonstrate both the importance of that which has been discussed and how the seemingly disparate pieces fit together to make a coherent whole.

Two-Dimensional Hydraulic Modeling for Alluvial Fan Floodplain Hazard Identification

Dragoslav L. Stefanovic* and Martin J. Teal[†]

West Consultants Inc.
11440 W. Bernardo Ct., Suite 360
San Diego, CA 92127
*dstefanovic@westconsultants.com
[†]mteal@westconsultants.com

Local and regional regulatory agencies are faced with the responsibility of ensuring that development on or near alluvial fans is carried out at an acceptable level of risk. This case study describes a project in Southern California which lies within an area of alluvial and flood deposit terrain northwest of the Borrego Sink in the Borrego Valley, San Diego County. The main purpose of this study was to determine the impact of proposed development on the flooding of the project site and adjacent areas for the 100-year flood event. A two-dimensional hydraulic and floodplain analysis was conducted as part of a proposed development. The two-dimensional hydraulic model FLO-2D was applied to analyze alluvial fan flood hazards, compute 100-year floodplain depths, and address on-site and off-site flow paths. The elements of this model and the modeling results for several design alternatives are presented here.

8.1 Introduction

Local and regional regulatory agencies are faced with the responsibility of ensuring that development on or near alluvial fans is carried out at an acceptable level of risk. Flood and debris hazard identification and mitigation are thus an important part of planned projects in these areas. This

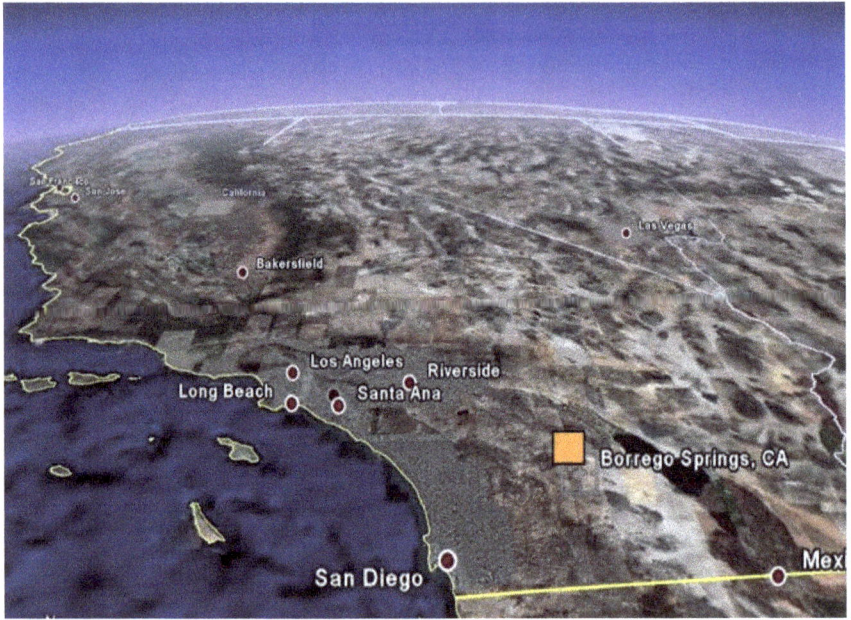

Figure 8.1   Project location map.

case study describes the analyses conducted as part of the Mesquite Trails
RV Park Development in Southern California. The project site lies within
an area of alluvial and flood deposit terrain northwest of the Borrego Sink
in the Borrego Valley, San Diego County, California, as shown in Figure 8.1.

The main purpose of this study was to determine the impact of the
proposed development on flooding of the project site and adjacent areas for
the 100-year flood. The project is limited to the westerly 40 percent [122
acres (49 hectacres)] of the 307-acre [124-hectacres (ha)] parcel (Figure 8.2)
in order to avoid construction in the existing Coyote Creek floodplain to
the north, denoted as the Fan Terminus Alluvial Wash (FTAW) on the
Federal Emergency Management Agency (FEMA) Flood Insurance Rate
Map (FIRM).

WEST Consultants (WEST) developed a two-dimensional FLO-2D
hydraulic model to analyze major flood hazards, compute 100-year floodplain depths, and address on-site and off-site flow paths. The modeling
approach developed is particularly effective for simulating unconfined flows
over complex alluvial fan topography and roughness. The elements of this

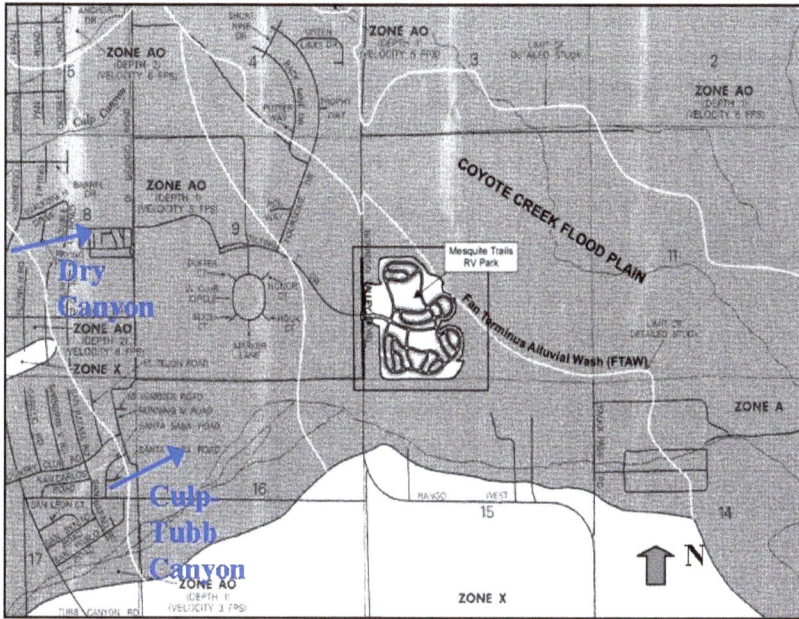

Figure 8.2   FEMA Flood Insurance Rate Map (FIRM) of the project area.

model and the modeling results for proposed design alternatives are presented in the following sections.

8.1.1   *Local regulatory framework*

In an effort to provide sustainable building practices following destructive flooding in 1976 and 1979, San Diego County authorized a report that was completed in 1989, the "Borrego Valley Flood Management Report" [Boyle Engineering, 1989]. The report contains the approved flood hazard map for the Borrego Valley alluvial fans, including the fan boundaries and depth-velocity lines. As shown on the FIRM, the FEMA design velocity in the project vicinity is 5.5 feet per second (fps) [1.5 meters per second (m/s)], the flow depth is 1.0 foot (0.3 m), and the energy grade line (flow depth plus velocity head) is 1.5 feet (0.5 m). The report also provides a design methodology for medium density developments using nonstructural methods, such that proposed construction will not cause a major alteration of the natural alluvial fan process. The report defines medium density developments as "projects designed so that flow can pass between obstructions and exit at the downstream end of the property in its natural condition without negative effects on the neighboring property." One of the major goals of the present analysis was to ensure that the project complies with those regulatory flood management requirements.

8.1.2   *Project setting*

The project site lies at the northeastern end of the Peninsular Range Batholith geomorphic province, an approximately 1,000-mile [1,609-kilometer (km)] long intrusive structure that extends from the transverse ranges through Southern California and throughout the length of Baja California in a south-southeast trending belt generally from 50 to 100 miles (80 to 160 km) in width. The site is situated between the Borrego Sink and a bajada formed along the northerly base of the Pinyon and Yaqui Ridges. Site area vegetation consists predominantly of mesquite, salt brush and perennial grasses on the easterly half of the property, and creosote bush and perennial grasses on the westerly half of the property. Soils exposed on the property consist predominantly of loamy coarse sand, sandy loam and silt loam.

Two major drainages flow through or adjacent to the property on their way to the Borrego Sink (a closed depression). The first drainage from Coyote Creek and its tributaries flows adjacent to the northeast corner of the property (Figure 8.2). The alluvial fan originating from the Coyote Creek watershed is the predominant geomorphic feature within the valley. The second drainage (Figure 8.3) originates from seven recognizable canyons that are incised into the relatively rugged terrain to the west; the largest are Culp Valley, Tubb Canyon, and Dry Canyon. Flow and debris exiting each canyon forms an alluvial fan, which often coalesces with an adjacent fan a short distance down into the valley. The alluvial fans from the west are slowly encroaching onto the Coyote Canyon fan, resulting in a significant grade break where the fans intersect. Within the Borrego Valley, the upper portions of the fans have surface gradients typically ranging from 4 to 8 percent, becoming progressively flatter down the fan. In the more developed portions of the valley, fan gradients range from 1 to 4 percent (the fan gradient at the project site is on the order of 1 percent).

Design floods in the Borrego Springs area are based upon summer tropical storms of brief duration and extreme intensity. It is indicative that average annual precipitation in the Borrego Valley is less than 6 inches

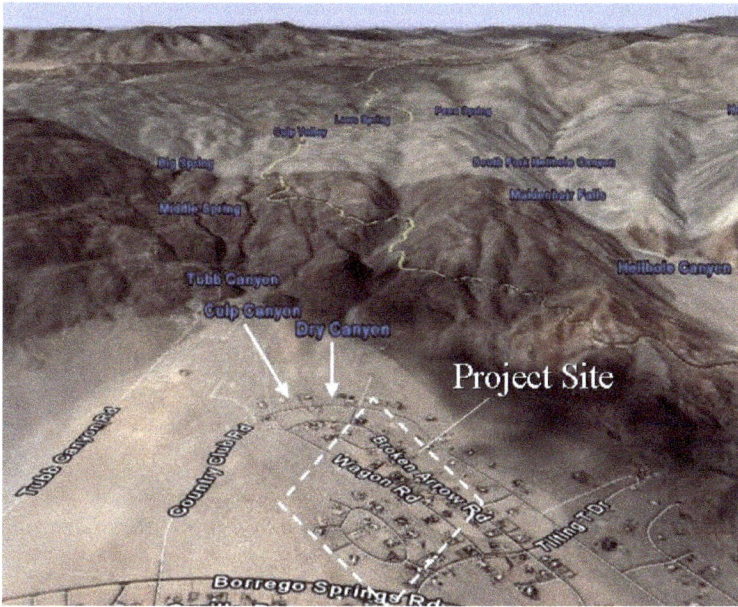

Figure 8.3   Alluvial fans west of project site.

[150 millimeters (mm)], while values along the western upstream ridge-line can reach 25 inches (640 mm), producing potentially destructive flash floods. Sizeable floods occurred in the past (1976 and 1979), and most of the flooding took place in the summer and fall months as a result of tropical cyclones generated in the South Pacific and Gulf of Mexico. During these types of severe storms, large quantities of overland flow develop in the sparsely vegetated granitic mountains that surround the Borrego Valley. Surface runoff collects in narrow steeply-walled canyons and, upon reaching the canyon mouth (fan apex), the unconfined water spreads into sheet flow across the wider alluvial surfaces.

8.1.3   *Hydraulic model development*

A two-dimensional hydraulic model FLO-2D [O'Brien, 2007] was prepared for the project site to study the alluvial fan flow from the west, mainly from the Culp-Tubb and Dry Canyon Fans (Figure 8.3). The study area encompasses some 365 acres (148 ha) between the FTAW boundary on the east and the 5.5/1.0 velocity-depth alluvial fan contour on the west. The model does not include flow from Coyote Creek, which enters the property on its northeast corner, because all residential development is located outside the FTAW boundary.

The original project design [GDC, 1993] incorporated a 400 feet (123 m) wide flood relief channel through the site that was proposed to accommodate the worst-case scenario flow of 9,000 cubic feet per second (cfs) [255 cubic meters per second (m^3/s)]. This hypothetical discharge was estimated by GDC as the combined apex flow from both Culp-Tubb Canyon and Dry Canyon, probabilistically much less than a 1 in 100 average annual occurrence at the site. However, the two-dimensional flow analysis performed by WEST identified that this single large channel located on the north side of the development was not effectively capturing actual drainage paths which generally follow the highest surface gradients. Also, the original lot alignment in the southern half of the site (Figure 8.4) was almost perpendicular to the flow from the west, obstructing the flow paths and causing excessive water ponding (more than 2 feet) (0.6 m) on the streets between the lots.

The original project design was revised, based on WEST recommendations, to generally align the south lots with the flow and direct flood waters

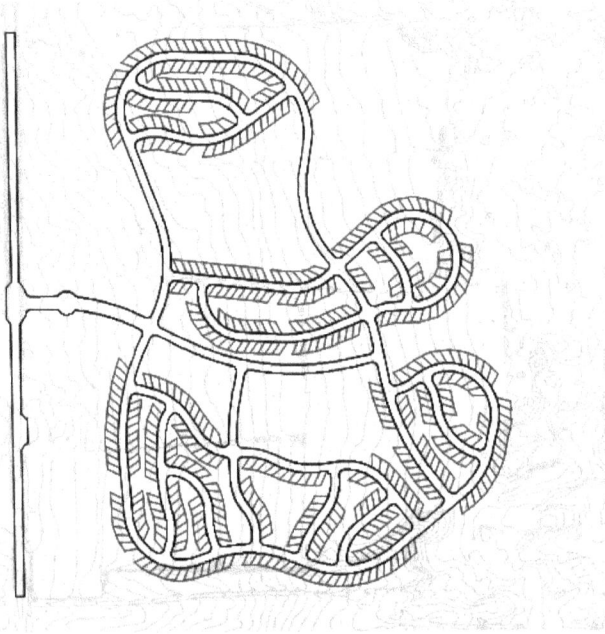

Figure 8.4   Original project design with lot layout.

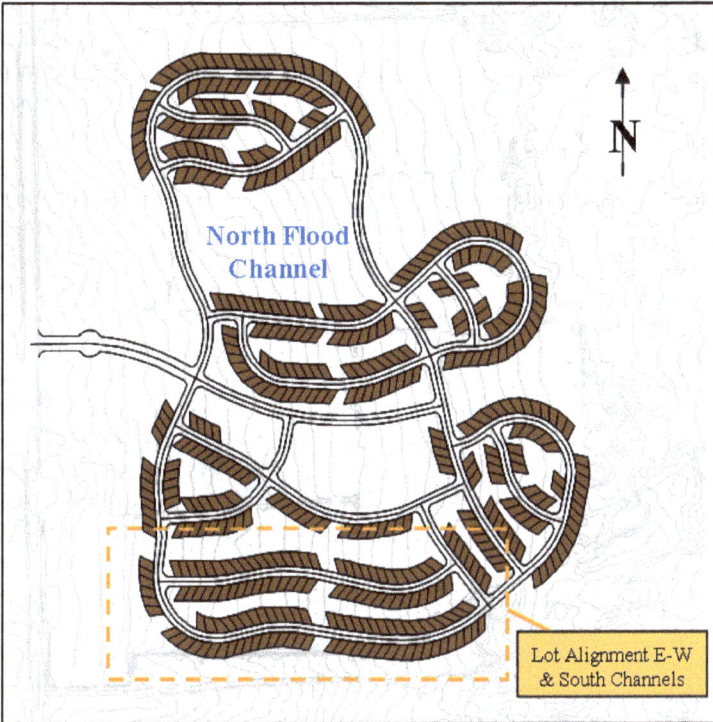

Figure 8.5   Revised project design with lot layout.

into channels between the lots (Figure 8.5) Also, the streets and pads on the south side were raised above the base flood elevations, while the channels between them were deepened to accommodate redistributed flood volumes. This project alternative enabled much better flow distribution through the site and reduced excessive flow velocities that were characteristic of the original design [up to 7 fps (2 m/s)].

In the final project alternative, the majority of south lots were shifted to the north, at the expense of the wide (and ineffective) original north flood relief channel (Figure 8.6). All of the southern channels were widened, which additionally decreased flood elevations in critical areas and significantly improved flow conveyance of the entire site. The multiple-channel layout proved to be far superior over the wide, single channel configuration because alluvial fan flows are prone to lateral migration and sudden relocation to any other portion of the fan during a single runoff event.

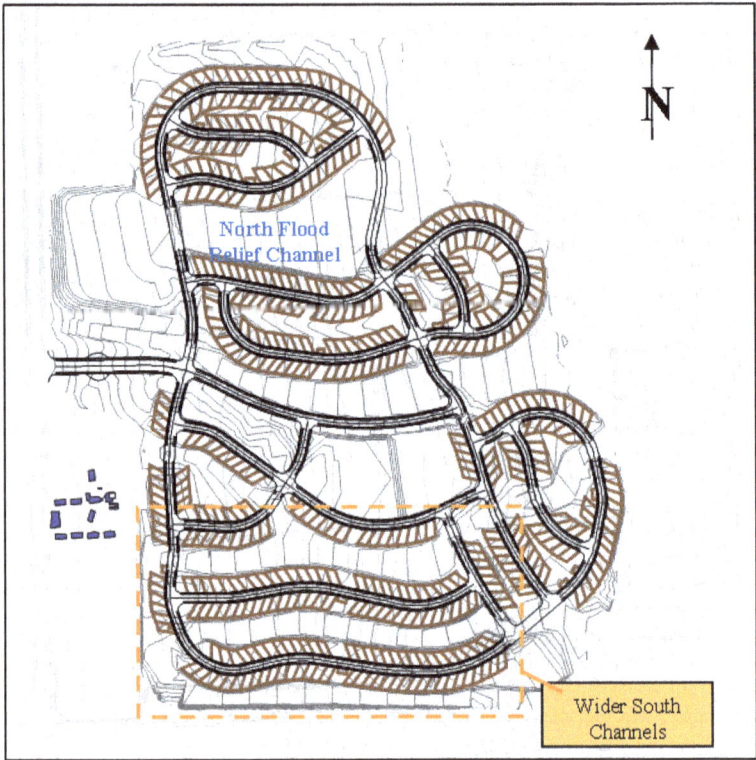

Figure 8.6 Final project design with lot layout.

8.2 Hydraulic Model Data and Assumptions

8.2.1 *Topography and grid development*

Two-foot (0.6 m) contour interval digital topographic data for the site were available from the owner. A digital raster graphic (DRG) based on a 10-foot (3-m) contour interval also was acquired from San Diego County to supplement the on-site topographic data and extend the terrain model beyond the project limits. The combined data were used to construct a Triangular Irregular Network (TIN) within the ArcView Geographic Information System program [ESRI, 2007]. The study area was divided into 6,300 square elements (50 by 50 feet) (15 by 15 m) within the FLO-2D grid developer system (GDS) and the TIN was imported in the background. Each grid element (center of the computational cell) was assigned a representative ground elevation from the TIN based on the interpolation algorithm embedded in the FLO-2D interface.

8.2.2 *Discharge*

Flooding on alluvial fans is considerably more complex than on conventional riverine systems. Flows rarely spread evenly over the surface of an alluvial fan. More typically, flow is concentrated in an identifiable temporary channel that can easily migrate to other areas of the fan surface. This erratic, unpredictable behavior subjects all portions of the fan to potential flood hazard, regardless of the location. Therefore, a site distant from an identifiable channel has approximately the same potential for flooding as a site at the same elevation (*i.e.*, the same radial distance from a fan apex) near an identifiable flow path.

Flood hazard assessment on alluvial fans incorporates this basic hydrologic perspective and assumes that a channel caused by a flood event is equally likely to cross a contour of the fan at any point on a specified contour. The flood insurance maps developed by FEMA assume that regime equations describe flow characteristics on the upper reaches of the alluvial fan and a finite width of flow will proceed down the fan depending on the probability that a given flood will flow past a particular point on the fan contour. This was the basis for the construction of depth-velocity contours in the Borrego Valley Flood Management Report [Boyle Engineering, 1989]. As noted in the report, these contours should not be construed as representing a flood flowing at that velocity and depth across the entire width of the fan. Instead, the flood would cut its own channel of a certain calculated width (based on the regime equation), flowing at the velocity and depth indicated on the flood hazard map. Each depth-velocity contour reflects the flow depth and velocity of a finite-width channel with an equal probability of crossing the contour at any given point (*i.e.*, the channel of finite width can migrate along the contour with an equal probability).

The Borrego Valley Flood Management Report indicates a design velocity, V, of 5.5 fps (1.7 m/s) and a specific energy (flow depth plus velocity head), H, of 1.5 feet (0.5 m) near the project site. The regime width of flow, W, using the Dawdy [1979] methodology is as follows:

$$W = 9.5Q^{0.4} \tag{8.1}$$

where Q is the channel discharge. This equation can be rewritten, solving for unit discharge q:

$$q = \frac{Q}{W} = 0.105Q^{0.6} \tag{8.2}$$

Therefore,

$$Q = [9.5q]^{1.67} \tag{8.3}$$

As indicated in the San Diego County's design criteria for nonstructural methods [Boyle Engineering, 1989], when using the FEMA flood hazard maps, critical flow is assumed and the flow depth, d, can be determined as:

$$d = \frac{2}{3}H = 1.0 \tag{8.4}$$

Since $q = Vd = 5.5$ cubic feet per second (cfs) $(0.2\,\text{m}^3/\text{s})$, the channel discharge from Eq. (8.3) is $Q = 740$ cfs $(21\,\text{m}^3/\text{s})$ and the corresponding regime flow width from Eq. (8.1) is $W = 133$ feet (41 m). This clear-water discharge is assumed to be the discharge in an individual channel of finite width [133 ft (41 m)] that would cross any point on the corresponding velocity-depth contour of the fan with an equal probability. The 100-year clear-water flow was eventually bulked to include the potential sediment volume carried by the flow (a minimum bulking factor of 1.5 was requested by the County of San Diego). The total (bulked) design discharge used in this study was 1,200 cfs $(34\,\text{m}^3/\text{s})$.

8.2.3 *Precipitation*

Based on precipitation data obtained from the National Oceanic and Atmospheric Administration Weather Service Office, the total 6-hour 100-year rainfall for Borrego Springs is 2.53 inches (64 mm) [NOAA, 2006]. A 6-hour rainfall hyetograph was developed based on methods outlined in the San Diego County Hydrology Manual [2003]. This 100-year rainfall pattern was uniformly distributed throughout the site and runoff was combined with the 100-year discharge to provide conservatively large flows (Figure 8.7).

8.2.4 *Infiltration*

Infiltration was simulated using the Green-Ampt model assigning uniform hydraulic conductivity and soil suction values throughout the development. The initial abstraction (filling of voids prior to simulating infiltration) was

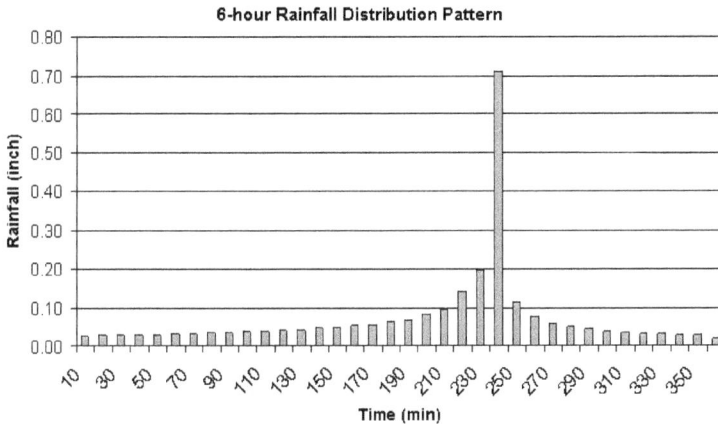

Figure 8.7   100-year precipitation pattern.

neglected for conservatism. A typical soil porosity of 0.4 was used. Based on the FLO-2D Manual [O'Brien, 2007], average hydraulic conductivity and soil suction were estimated at 0.8 in/hr (20 mm/hr) and 3.3 in (84 mm), respectively, for the soil material present at the site (silty sand to sandy silt). Streets and lots were modeled as impervious surfaces.

8.2.5   *Manning's n-values*

The assignment of overland flow resistance must account for vegetation, surface irregularity, flow depth, and flow path redirection. Overland roughness values can be two or three times those used for conventional open channel flow. Therefore, Manning's roughness values were carefully selected based on field observations and engineering judgment, with guidance from the FLO-2D Manual [O'Brien, 2007]. Table 8.1 provides a summary of the roughness n-values used for this study.

Table 8.1   Manning's n-values.

Type of channel or overland area	Manning's n-value
Sparsely vegetated, uniform 400' wide drainage channel	0.050
Floodplain (mesquite, salt brush, perennial grass)	0.070
Dense trees around the existing houses on the west	0.120
Shallow overland flow (depth < 0.2 feet)	0.200
Streets and roads	0.025

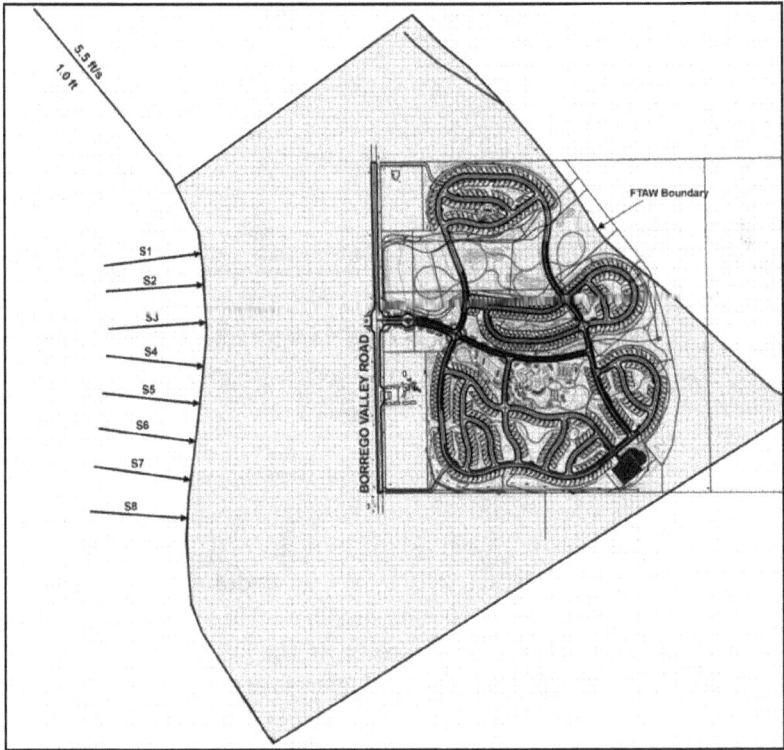

Figure 8.8 Inflow points and computational domain. On the downstream (east) end (FTAW Boundary), the normal depth outflow condition (with a friction slope equal to the surface gradient) was used. The assumption of normal depth was deemed reasonable because the alluvial fan flow from the west will not be obstructed by Coyote Creek flow along the FTAW boundary. (The floodplain limits are shown in Figure 8.1.)

8.2.6 *Boundary conditions*

FLO-2D requires that an inflow hydrograph be specified at the upstream end of the computational domain and an outflow condition at its downstream end. The upstream (west) boundary in this study was conveniently selected to coincide with the 5.5/1.0 velocity-depth alluvial fan contour, while the downstream (east) boundary coincides with the FTAW boundary (Figure 8.8).

Because the temporary alluvial fan channel [with a regime width of 133 ft (41 m) and bulked discharge of 1,200 cfs (34 m^3/s)] is equally likely to cross the contour of the fan at any point along the contour, the inflow hydrograph was successively specified at eight locations (S1 through S8 in Figure 8.8) that approximately cover the entire west boundary of the domain.

Each flooding scenario was simulated in turn, and the highest flow depth and maximum velocity among the eight scenarios at each computational element were selected to represent the most critical flooding condition. Because the regime flow width is approximately 133 ft (41 m), three computational cells [a total width of 150 ft (46 m)], with 400 cfs (11 m^3/s) per cell, were used to define the total bulked inflow as the upstream boundary condition in each scenario.

8.2.7 Flow obstruction

One of the unique features of the FLO-2D model is its ability to simulate problems associated with flow obstructions. Area reduction factors (ARFs) and width reduction factors (WRFs) are coefficients used to modify the individual grid element surface area storage and flow width. These factors greatly enhance the detail of the flood simulation through an urbanized area. The ARFs were used in this study to reduce the flood volume storage on grid elements due to building lots (Figure 8.9).

The WRFs were assigned to particular flow directions in a grid element to refine the flow around the lots and down the streets. The split rail fence (Figure 8.9, dashed line) also was included along the west and east side of the development, following the FTAW boundary in the northeast quadrant. A flow width reduction of 30 percent (WRF = 0.7) was assumed in the computational cells along the fence to simulate flow blocking.

8.2.8 Froude number

Establishing a limiting Froude number in a flood routing model helps sustain the numerical stability by forcing the model to have a reasonable representation of physical reality. Overland flow on steep alluvial fans often approaches critical flow. In general, supercritical flow on alluvial fans is suppressed by high rates of sediment transport, hence high velocities and shallow depths on alluvial surfaces will dissipate energy with sediment entrainment (supercritical flow is more prevalent on bedrock or other hard surfaces). Therefore, a limiting Froude number of 0.9 to 0.98 is recommended for an alluvial fan analysis. In this study, a limiting Froude number of 0.9 was found to provide stable results.

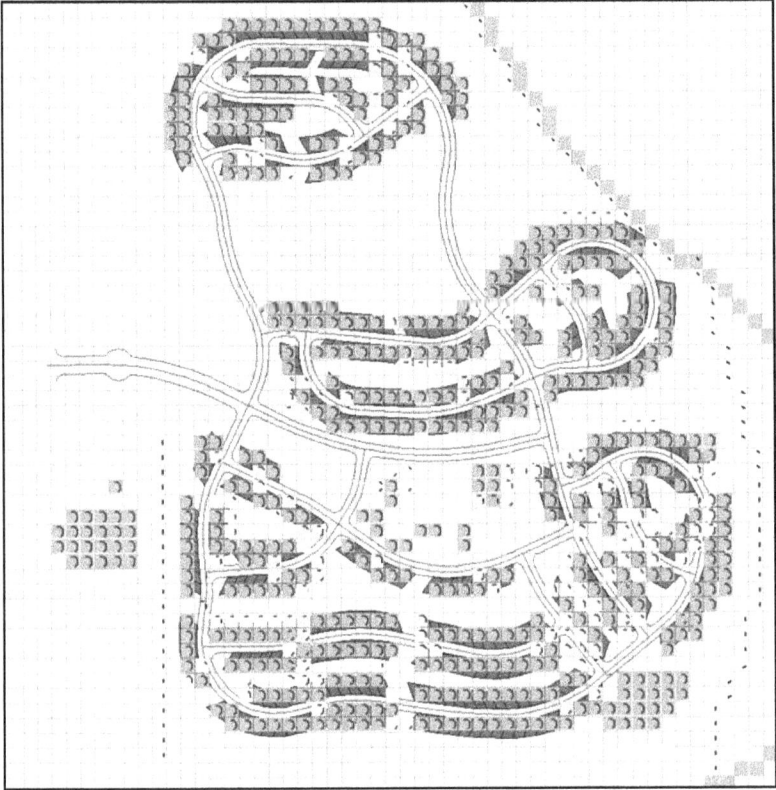

Figure 8.9   Blocked computational domain cells with ARF and WRF reduction factors.

8.2.9   *Computational time step and grid element size*

FLO-2D is an explicit finite difference model that requires careful selection of the computational time step to maintain numerical stability. This selection process is accomplished internally by the program based on the Courant stability criterion which ties the computational time step to the grid element size. In order to prevent the model from running exceedingly slow, it is recommended to select the grid element surface area, A, such that $Q/A < 0.5\,\text{cfs}/\text{ft}^2$ $(0.15\,\text{m}^3/\text{s}/\text{m}^2)$ where Q is the peak discharge in the grid element. Based on this criterion, the grid element size 50 by 50 feet (15 by 15 m) was selected $(Q/A = 0.16)$. This grid size was found to generate a reasonable time step (on the order of a few seconds) and provide enough spatial resolution to describe critical floodplain elements.

8.3   Hydraulic Model Results

Each flooding scenario described in Section 8.2.6 was simulated for six hours of constant alluvial fan inflow [1,200 cfs (34 m^3/s)] plus rainfall, which was sufficient to develop steady-state conditions (*i.e.*, no discernible change in depth and flow velocity thereafter) on the floodplain.

Maximum flow depth and velocity fields for the original project design are shown in Figure 8.10. Most of the flow is directed east-west except in the southwest portion of the development where the inadequate lot orientation diverts the flow in the north-south direction down the streets. This created several stagnation points with relatively high depths on the streets [above 2 ft (0.6 m)], alternated by self-accelerating flows with maximum velocities on the order of 7 fps (2 m/s). Both conditions were unacceptable because the project requirements (specified by the San Diego County Flood Control District) were to limit the street flow depths below 1.0 ft (0.3 m) and maximum velocities below 5.5 fps (1.7 m/s).

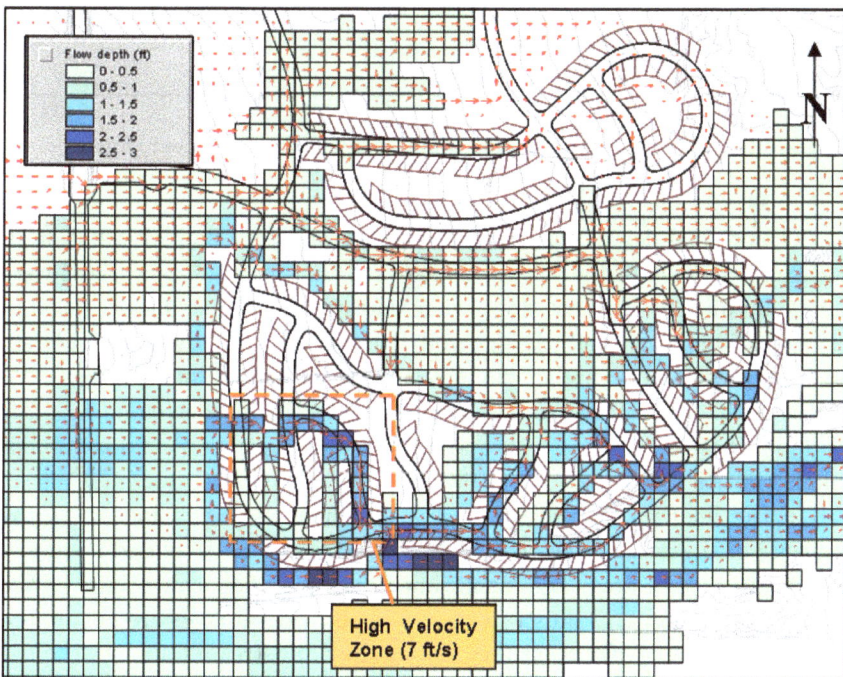

Figure 8.10   Simulated maximum flow depths and velocities for the original project design.

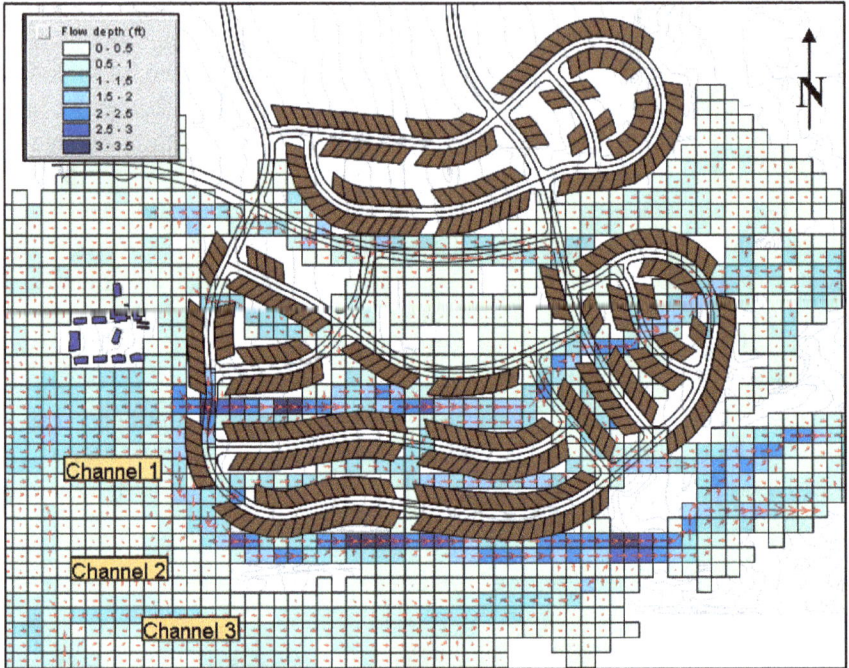

Figure 8.11   Simulated maximum flow depths and velocities for revised project design.

The original design was revised to allow for better flow distribution in the south portion of the site as discussed in Section 8.1.3. South lots were generally aligned with the flow, while the streets and pads were elevated in a few iterations until the alluvial fan (off-site) flows were completely diverted to three flood relief channels between the south lots (Figure 8.11). As a result, the maximum flow depths on the streets decreased to 0.3 ft (0.1 m) (these depths are only due to on-site precipitation) and the maximum velocities were reduced below 1 fps (0.3 m/s).

In the final project design, the south channels were widened at the expense of the north relief channel, which additionally decreased flood elevations in critical areas and significantly improved flow conveyance of the entire site. Figure 8.12 illustrates how the alluvial fan flows are now unobstructed by the development and exit at the downstream (east) end of the project without negative effects on the surrounding properties. The proposed multiple-channel approach also helped to better balance the site in terms of cut and fill quantities for lot construction.

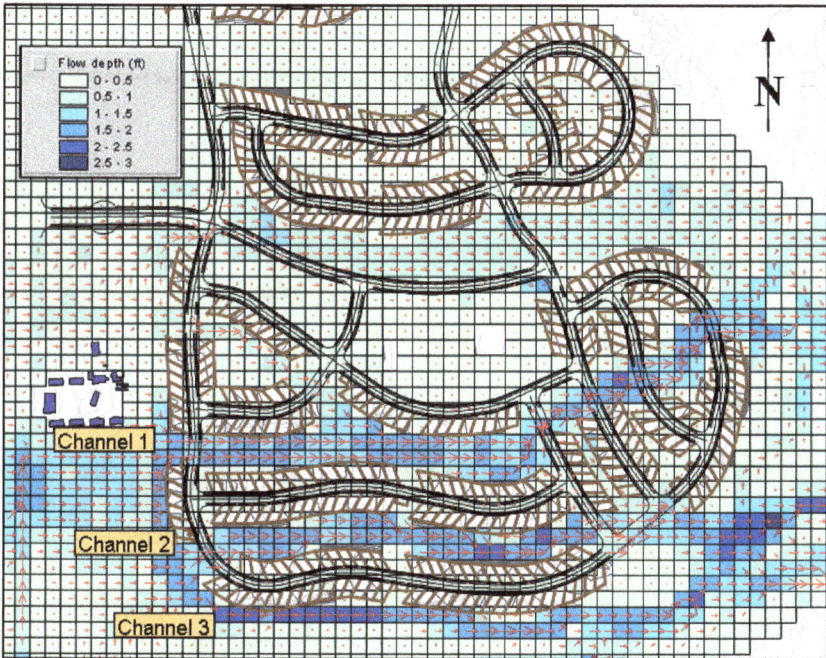

Figure 8.12   Simulated maximum depths and velocities for the final project design.

8.4   Summary and Conclusions

Flooding on alluvial fans is considerably more complex than on conventional riverine systems. Typically, flow is concentrated in an identifiable temporary channel that can easily migrate to other areas of the fan surface. This erratic, unpredictable behavior subjects all portions of the fan to potential flood hazard, regardless of the location. Consequently, a multi-dimensional hydraulic analysis is generally preferred over a one-dimensional approach for alluvial fan floodplain hazard identification.

WEST prepared a two-dimensional (FLO-2D) hydraulic model to analyze alluvial fan flood hazards for a typical development in Southern California. The model was utilized to predict maximum floodplain depths, and address off-site and on-site flow paths due to 100-year alluvial fan flooding combined with 100-year on-site precipitation. As a result of comprehensive model simulations, the project site was designed to be capable of accommodating the alluvial fan flows by allowing them to pass through the property without causing significant flow obstruction.

.

The WEST study approach combines an innovative modeling technique with a design methodology suitable for site development on alluvial fan floodplains using nonstructural methods. The main features of this unique approach and resulting design guidelines are the following:

(1) Because the temporary alluvial fan channel of finite width is equally likely to cross the contour of the fan at any point, the flooding source is successively specified at multiple locations along the entire inflow boundary of the domain. Each flooding scenario, pertaining to one flooding source location, is simulated in turn. The highest flow depth and maximum velocity among the multiple scenarios at each computational element (cell) are selected to represent the most critical flooding condition.

(2) The least flow obstruction is accomplished if the structures on the site (buildings, roads, *etc.*) are generally aligned with the flow. This prevents large stagnation areas where the flow significantly slows down, causing excessive flow depths and potential sediment/debris accumulation.

(3) The best flow distribution (site conveyance) is provided using multiple flood relief channels between lots. This layout is preferred over a wide, single channel configuration because alluvial fan flows are prone to sudden lateral migration (one channel may not be able to capture all drainage paths that generally follow the highest terrain gradients). It may also be favorable to raise the streets above the base flood elevations, while redirecting all the off-site flows into flood relief channels (this way streets will convey only surface runoff).

(4) The multiple-channel approach generally helps to better balance the site in terms of cut and fill quantities for the lot construction, thus reducing the overall project cost.

References

Boyle Engineering Corporation (1989). Borrego Valley Flood Management Report, prepared for County of San Diego.

Dawdy, D.R. (1979). "Flood Frequency Estimates on Alluvial Fans," J. of the Hydraulics Division, ASCE, Vol. 105, No. HY11, pp. 1047–1413.

ESRI (2007). ArcGIS 9.0, Redlands, California.

Group Delta Consultants — GDC (1993). Limited Geotechnical and Hydrologic Investigation for Draft Environmental Impact Report, Mesquite Trails RV Park, Borrego Springs, California.

National Oceanic and Atmospheric Administration (NOAA) (2006). Atlas 14 Precipitation-Frequency Atlas of the United States, Version 4: Borrego Desert Park, Southern California. Silver Spring, Maryland, (http://hdsc.nws.noaa.gov).

O'Brien, J. (2007). FLO-2D Users' Manual, Version 2007.06, Nutrioso, AZ.

San Diego County (2003). Hydrology Manual. County of San Diego Department of Public Works, Flood Control Section.

Case Study #2

Numerical Modeling of the 2005 La Conchita Landslide, Ventura County, California

Philip J. Shaller,* Parmeshwar L. Shrestha, Macan Doroudian, David W. Sykora and Douglas L. Hamilton

Exponent, Inc.
320 Goddard, Suite 200, Irvine, CA 92618
*pshaller@exponent.com

Large-scale debris flows represent rare but destructive geologic hazards in the western United States. One such event took place on January 10, 2005 in La Conchita, California. Triggered by intense rainfall, coupled with unfavorable geologic conditions, the initial slope failure mobilized over $40,000 \, yd^3$ ($30,000 \, m^3$) of debris, sending a wall of wet soil and broken rock into a residential community, killing 10 persons and damaging or destroying 36 residences. Following both common and technical usage for this event, the name "La Conchita landslide" is used throughout this chapter to describe both the initial slope failure and the subsequent flow of wet debris into the community. As part of a causation analysis, the flow of debris was modeled using the commercial FLO-2D software package. The FLO-2D analyses offered a good approximation of the actual behavior of the La Conchita landslide, as validated by means of aerial photographs, a news crew video of the event, and field investigations. This investigation demonstrates the profound benefits gained by utilizing a multi-disciplinary approach to problem solving and harnessing recent advances in computing power, software and processing techniques.

8.5 Introduction

The small community of La Conchita is located along the picturesque coastline between Los Angeles and Santa Barbara, California. Although the region is renowned for its mild climate, powerful winter storm sequences pummel the region at least once each decade. In late 2004 and early 2005, an intense series of rainstorms impacted southern California, causing flooding and innumerable landslides throughout the region. The January 10, 2005 La Conchita landslide was the deadliest single event triggered by the 2004–2005 storm sequence.

Debris flows are an important class of mass movement in arid and semi-arid environments in the western United States. From a geomorphic perspective, debris flows are significant agents of erosion in mountainous watersheds and a principal mechanism by which coarse-grained sediment is transported onto alluvial fans. They also pose a significant hazard to residents and infrastructure in the region.

Debris flows are a particular hazard in the western United States due to the confluence of several regional factors, including:

(1) active tectonics, which create rugged terrain, tectonically disturbed source materials, and groundwater anomalies;
(2) a seasonal climate, in which most of the precipitation occurs over a relatively short time, either during the winter or during a summer monsoon;
(3) dynamic wildfire conditions, which, when coupled with seasonal rainfall, result in the "fire-flood-erosion sequence," in which debris flows play a major role;
(4) an extensive, distributed infrastructure system that is susceptible to damage by debris flows; and,
(5) a locally dense and rapidly growing population.

Several of these factors were involved in the case of the La Conchita landslide.

Large-scale debris flows are relatively uncommon. Nevertheless, they are profound agents of geomorphic modification as well as major geologic hazards. They also offer a unique window into the mechanics of debris flows. This is because their sheer size subordinates the effects of other factors, such as channel sinuosity, ground roughness, localized topography and vegetation cover.

The work reported in this chapter reflects broad trends affecting a range of sciences today, utilizing a multi-disciplinary approach to problem solving, and exploiting advances in computer modeling and processing capabilities. This investigation involved a close collaboration between geology, surface water hydrology, and geotechnical engineering practitioners. Moreover, the intensive modeling efforts described herein required processing capabilities that were, until recently, confined to large, dedicated main-frame computing centers. Hopefully, this methodology can be applied elsewhere to help recognize and mitigate the hazards presented by debris flows and lessen the potential for future tragedies such as that which struck La Conchita on January 10, 2005.

8.6 Background, Geology, and Kinematics

8.6.1 *Introduction*

The January 10, 2005 La Conchita landslide occurred along a sparsely developed coastal region about 81 mi (130 km) northwest of downtown Los Angeles, California, at *latitude 34.36 degrees N, longitude 119.45 degrees W.* The slope failure occurred at the head of a shallow swale located along the eastern lateral margin of a prior slump failure that affected a much larger portion of the slope in 1995 (Figure 8.13).

The 2004–2005 winter rainfall season was marked by a series of major Pacific storms that brought heavy precipitation to California. The first major Pacific storm of the season occurred in October. Heavy rainfall then returned on December 27th. The storms took a very consistent track colloquially known as the "Pineapple Express." During this storm sequence, Los Angeles had its wettest 15-day period on record. Between January 6th and 11th, over 20 in (500 mm) of rainfall was recorded at mountain weather stations in Santa Barbara, Ventura and Los Angeles Counties [NCDC, 2005]. Approximately 14.9 in (378 mm) of rain fell in the City of Ventura between December 27th and January 10th, only slightly less than the 15.4 in (390 mm) average yearly rainfall for the region [Schiek and Hurtado, 2005].

Material from the initial slope failure rapidly transformed into a large-scale debris flow. The majority of the debris flow traveled down the swale between the 1995 slump and the adjacent hillside area, curved to the left (looking downhill) following the pre-existing topography, and entered the residential community of La Conchita. This lobe of the debris flow traveled about 330 ft (100 m) through the community, destroying approximately 30 homes and resulting in the deaths of 10 persons [Gurrola, 2005]. A second lobe of the debris flow, containing about 10 percent of the total mobilized volume of debris, traveled down the heavily vegetated slope to the west of the main lobe. This minor lobe impacted and partially breached the central part of a temporary retaining wall located along Vista del Rincon Avenue

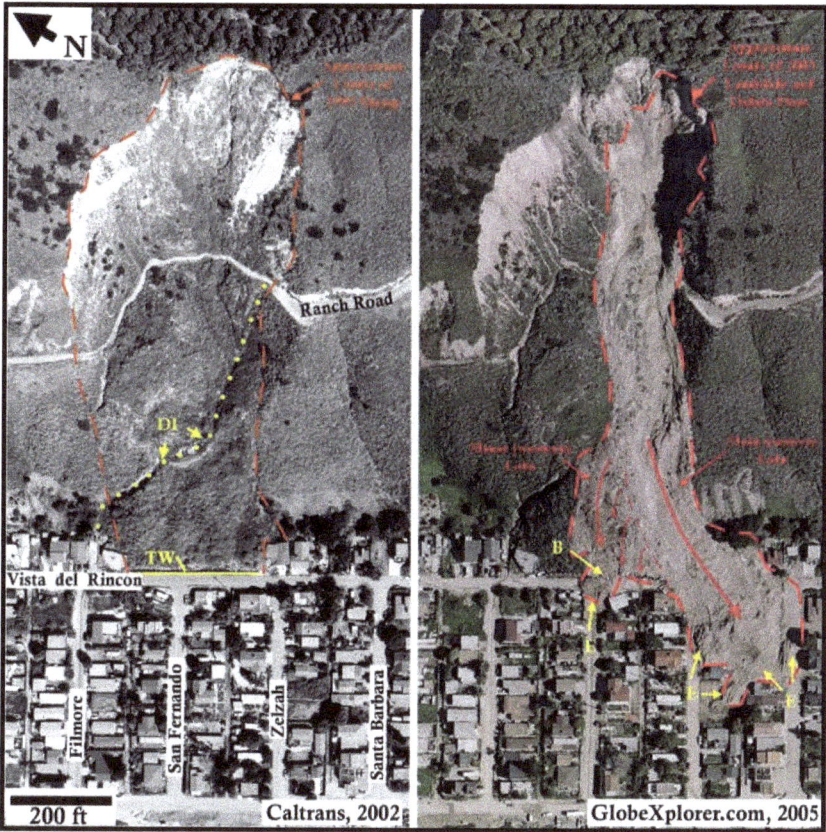

Figure 8.13   Comparison of aerial photos of the La Conchita area taken in 2002 (left) and 2005 (right). The photo at left shows the approximate limits of the 1995 slump failure, the location of a drainage improvements (DI) made to convey water across the slump failure to the area above Fillmore Avenue (yellow dotted line), and the location of a temporary wall (TW) constructed along the uphill side of Vista del Rincon Avenue to allow reopening of the street in 2000 (solid yellow line). The photo at right, taken shortly after the 2005 landslide, defines the main (eastern) and minor (western) lobes, the inferred movement direction of the debris in these lobes, and the location of a breach (B) in the temporary wall caused by impact of the debris, as well as locations affected by emergency grading after the event (E).

(see Figure 8.13), but caused no additional personal property damage or loss of life in the community. Remarkably, a television news crew that was in the area to cover earlier rain-related highway and rail line closures videotaped the incipient failure and much of the debris flow as it descended the slope.

8.6.2 *Historical setting*

The La Conchita area has a long history of development and landslide activity. The Southern Pacific Railroad laid rail lines through the area in 1887. By 1889, sections of the tracks had already been impacted by landslide debris. The tracks were moved in 1909 after a work train was buried in a landslide. That year, in an effort to reduce the hazard posed by the steep bluff face, the railroad leveled the area between the tracks and the slope. The La Conchita residential development was established on the leveled area in 1924 [Hemphill, 2001].

The bluff and hilltop terrace areas upslope from La Conchita formed a portion of the La Conchita Ranch property (Ranch). The Ranch originally practiced grazing and dry agriculture, then switched to irrigated agriculture in the mid-1970s [Hemphill, 2001]. In March 1995, a large slump failure occurred on the bluff following a period of heavy rainfall. This event damaged or destroyed seven residences, destroyed a portion of Ranch Road, an access road that traversed the bluff face, and covered a major street in the community, Vista Del Rincon Avenue, with up to about 20 ft (6 m) of debris (Figure 8.13). Despite its large volume, estimated at between 260,000 and 600,000 yd^3 (200,000 and 460,000 m^3), the landslide was anticipated and slow moving, allowing residents time to evacuate [O'Tousa, 1995; Harp *et al.*, 1995].

The slump failure was never remediated. By early 1996, however, the Ranch had performed minor earthwork to convey stormwater from the truncated uphill portion of Ranch Road across the slump to a point along the toe of the 1995 slump at the northern end of Fillmore Avenue. These drainage improvements (DI) are indicated on Figure 8.13. Also shown in the figure is a steep-sided channel incised by surface runoff into the western side of the slump by 2002.

In 2000, the County of Ventura constructed a temporary soldier pile wall consisting of steel H-piles and wood lagging wall along the northern margin of Vista del Rincon Avenue to allow the removal of debris from the street. The wall was 270 ft (82 m) long, stood between 5 to 23.5 ft (1.5 to 7.2 m) above the road surface (including a guardrail at the top) and had a freeboard of about 4 to 9 ft (1.2 to 2.7 m), including the guardrail.

Litigation triggered by the 2005 debris flow focused in part on the role played by the temporary wall in affecting the path taken by the flow as it entered the community. The FLO-2D investigation described herein sought to quantify the effects of the wall on the movement rate, travel direction, and area inundated by the debris flow.

8.6.3 *Geologic conditions*

The 2005 slope failure originated in old landslide deposits near the crest of a 590 ft (180 m) high, southwest-facing coastal bluff. The bluff represents a modified Holocene sea cliff that is capped by the 45,000 yr BP Punta Gorda marine terrace. The Punta Gorda terrace formed when sea level was about 125 ft (38 m) below present-day sea level, indicating that the terrace is rising at a long-term geologic rate of over 0.16 in/yr (4 mm/yr) [Huftile *et al.*, 1997].

As shown on Figure 8.14, most of the bluff above La Conchita is mantled by landslide deposits. These landslides occurred within a sequence of poorly indurated sedimentary rocks of Tertiary age. The principal geologic units in the vicinity include the Upper Middle Miocene Monterey Formation (shale, siltstone, and sandstone), the Miocene-Pliocene Sisquoc Shale (silty shale and claystone) and the Pliocene Pico Formation (sandstone and conglomerate). The distribution of bedrock units in the bluff is obscured by the thick mantle of landslide deposits and complicated by the presence of the Red Mountain fault, which is mapped intersecting the slope face in the source area of both the 1995 and 2005 landslides (Figure 8.14).

The Red Mountain fault is a major, active reverse fault. Huftile *et al.* [1997] observed 112 ft (34 m) of vertical separation of the Punta Gorda wave-cut platform along the fault, corresponding to a dip-slip rate of about 0.06 in/yr (1.5 mm/yr). Figure 8.15 shows an interpretation of the subsurface conditions underlying the bluff above the community of La Conchita, including the inferred location and orientation of the Red Mountain fault, offset bedrock units, preexisting landslide deposits, and the Punta Gorda marine terrace [Rogers *et al.*, 2007].

8.6.4 *Vegetation and soils*

At the time of the 2005 landslide, thick chaparral vegetation mantled the face of the bluff and the surface of the 1995 slump. The terraced area at the top of the bluff was developed by the Ranch as an avocado orchard. The surficial soils in the La Conchita area consist of fine-grained silt loam and shale loam soils [USDA, 2008].

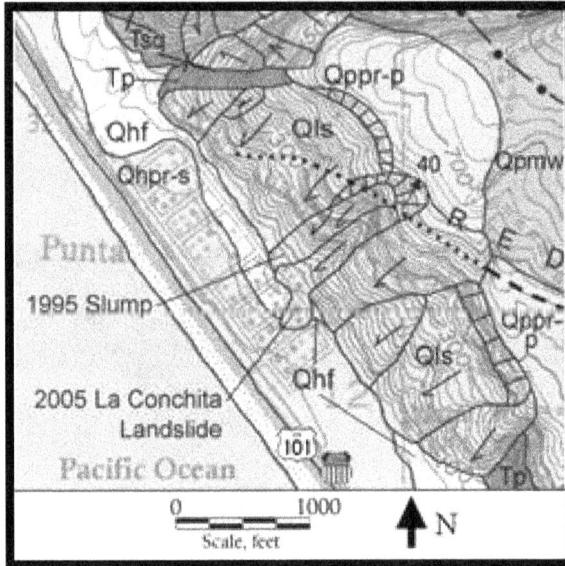

Figure 8.14 Geologic map of the 2005 La Conchita landslide and vicinity. The Red Mountain fault extends through the headscarp area of the 2005 slope failure and the earlier 1995 slump failure. Geologic unit designations: Qls — Landslide deposits; Qhf — Undivided alluvial and colluvial deposits; Qhpr-s — Terrace deposits associated with 1,800–5,800 BP Sea Cliff marine terrace; Qppr-p — Terrace deposits associated with 40,000–60,000 BP Punta Gorda marine terrace; Qpmw — Undivided Pleistocene talus, colluvium and landslide deposits; Tp — Pliocene Pico Formation; Tsq — Miocene-Pliocene Sisquoc Shale. Dash-dot line indicates inferred shoreline angle of Punta Gorda terrace. Modified from Geologic Map of the Pitas Point 7.5-Minute Quadrangle, USGS, 2005.

The principal morphological elements of the 2005 La Conchita landslide are illustrated on Figure 8.16, a ground photo taken on the day of the event. The main (eastern) lobe of the deposit exhibits raised lateral levees, a common feature of small scale mudflows and debris flows [Sharp and Nobles, 1953; Johnson, 1970]. It also exhibits a ropy pattern of ridges and troughs aligned transverse to the direction of movement. This ropy pattern is similar in appearance (but not scale) to the morphology that commonly develops on more fluid lava flows (pahoehoe), which have also been characterized as natural plastics [Hulme, 1974]. A similar surface texture has also been described from a very large debris flow deposit in central Idaho [Shaller, 1991a] and from several giant Martian landslides [Shaller, 1991b].

The minor lobe of the La Conchita deposit (to the right in Figure 8.16) lacks these morphological characteristics, instead exhibiting a hummocky,

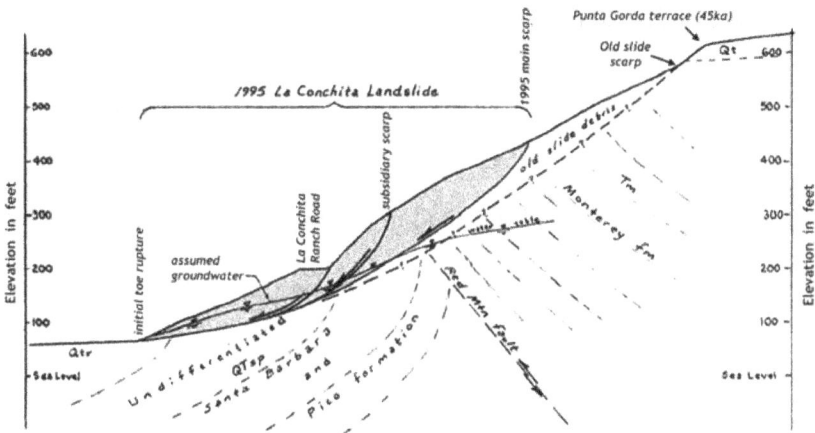

Figure 8.15 Conceptual geologic cross section through the bluff face in the vicinity of the 1995 and 2005 La Conchita landslides [modified from Rogers *et al.*, 2007].

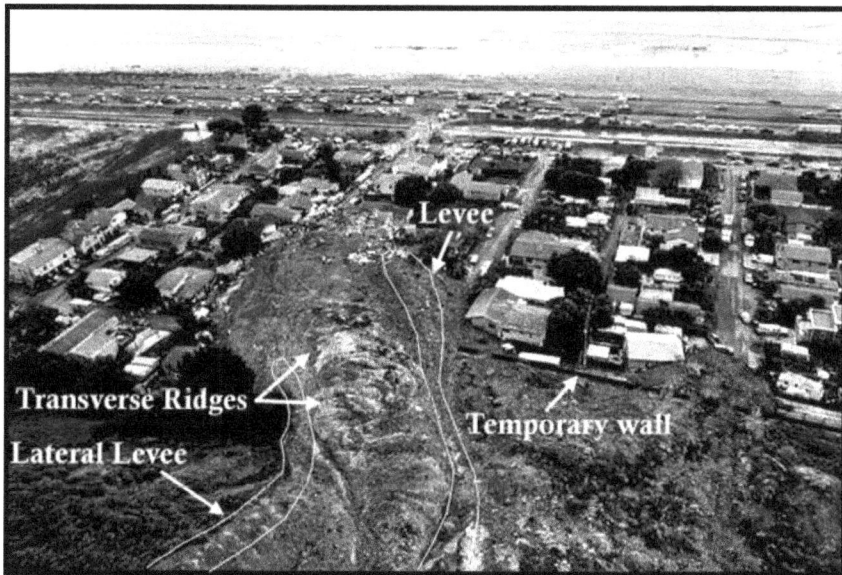

Figure 8.16 Oblique ground photo of the La Conchita landslide taken shortly after the event. The lateral margins of the main lobe are marked by moderately well developed levees (circled). The interior of the deposit exhibits a ropey pattern of ridges and troughs aligned transverse to the travel direction. The location of the temporary wall constructed to reopen Vista del Rincon Avenue is indicated in right center of image. The dashed portion of the line indicates portions of the wall that were breached or otherwise covered with debris following the event.

irregular surface texture. The contrast in surface morphologies between the two lobes implies differences in their emplacement mechanisms. The geometry and morphology of the minor lobe suggest it formed by way of two processes:

(1) fluid debris entering the Ranch Road drainage channel, then overflowing onto the slope face; and,

(2) debris from the western edge of the main (easterly) lobe overflowing the swale as the result of toboggan like superelevation as it banked to the left in the mid-slope area on its way down slope (Figure 8.13).

Notably, no morphological evidence exists to suggest that the temporary wall erected by the County in 2000 deflected or otherwise significantly altered the direction of the debris flow prior to entering the community. As shown on Figure 8.16, the levee bounding the western (right) side of the main lobe shows no significant deflections where it crosses the eastern margin of the temporary wall. This observation is consistent with the behavior of large, rapid landslides elsewhere. Due to their substantial thickness and momentum, these landslides rarely demonstrate any significant deflections unless the object encountered is of sufficient height and strength to avoid being overridden, crushed or displaced by the onrushing debris. Excellent examples of the interaction of large-scale landslides with obstacles in their path are available in a series of photos of large, rapid landslides triggered by the 2002 Denali earthquake that traveled across Black Glacier, Alaska [USGS, 2008]. Many of these images show the landslide debris draped over ~ 30 ft (~ 10 m) high medial moraines located across their path.

8.6.5 *Sedimentology*

Test pits were excavated into the minor lobe of the La Conchita landslide behind the temporary wall. The debris exposed in these test pits consisted of pale yellow-brown, low plasticity silt exhibiting a massive, ungraded texture. A key observation in the test pit exposures was the presence of slope-parallel layers of pulverized vegetation located at approximately 4 ft (1.2 m) vertical intervals. This evidence indicates that the minor lobe was deposited in discrete pulses.

Table 8.2   Key physical measurements of the January 10, 2005
La Conchita landslide.

Volume — upper depletion zone	27,190 m^3	35,536 yd^3
Volume — lower depletion zone	4,140 m^3	5,411 yd^3
Total depletion volume	31,330 m^3	40,947 yd^3
Accumulation volume	27,090 m^3	35,406 yd^3
Estimated removals	4,240 m^3	5,541 yd^3
Average thickness	1.8 m	6 ft
Length (L)	407 m	1,335 ft
Maximum fall height (H)	152 m	500 ft
H/L (Fahrböschung)	0.37	0.37
$\tan^{-1}$(H/L)	20.5°	20.5°
Maximum width	76 m	250 ft
Velocity — slope area	~6–10 m/s	~13–22 mi/hr
Velocity — community	~5 m/s	~11 mi/hr

8.6.6   *Physical dimensions*

Key physical measurements of the January 10, 2005 La Conchita landslide are reported in Table 8.2 and illustrated in Figure 8.17. Figure 8.17 shows differences in elevation (feet) between 2002 (pre-event) and 2006 (post-event) topographic surveys of the area. The individual topographic surveys were developed using photogrammetric methods from high-resolution vertical aerial photos. Red shading indicates areas where the elevation increased between the two surveys (zone of accumulation); blue areas indicate areas where the elevation decreased between the two surveys (zones of depletion). The gray area near the toe of the deposit indicates areas modified by grading during the emergency response immediately following the disaster.

The volume of the initial slope failure was approximately 35,536 yd^3 (27,190 m^3), corresponding to the upper area of depletion located between elevation 250 and 530 ft (76 and 162 m; Table 8.2, Figure 8.17). Geomorphic evidence and the available video footage indicate that, once initiated, the landslide rapidly transformed into a large-scale debris flow. As it traveled downslope, the main (eastern) lobe eroded and entrained material along its path, creating the lower zone of depletion shown on Figure 8.17. This conclusion is based on the following observations:

(1) The lower depletion zone was located directly in the path of the main lobe;
(2) The material occupying the swale was likely in a wet, easily erodible condition due to the heavy antecedent rainfall; and,

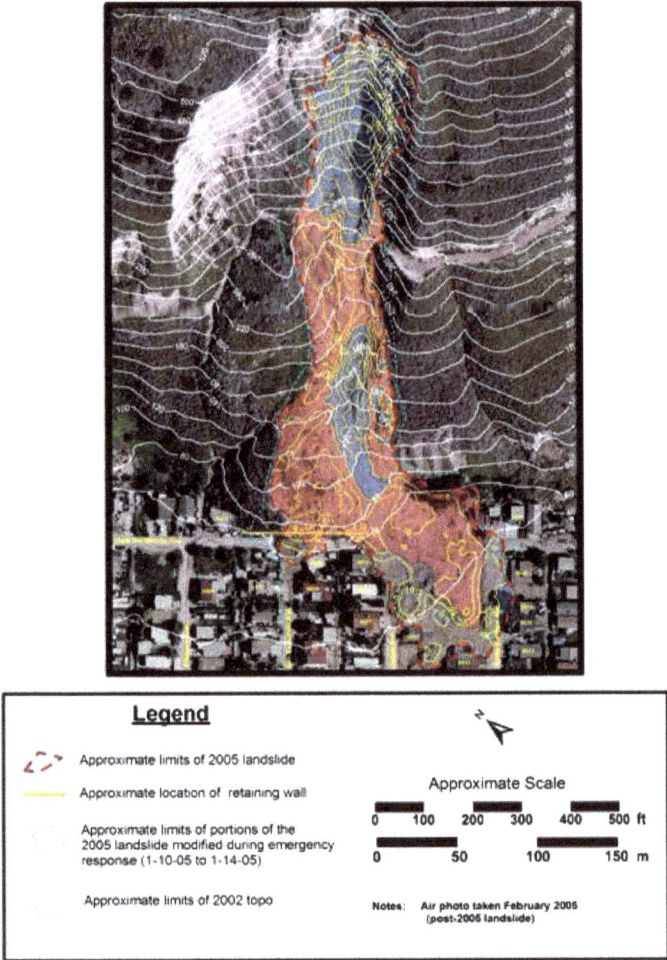

Figure 8.17 Topographic map of the January 10, 2005 La Conchita landslide showing change (feet) between elevations reported on 2002 (pre-event) and 2006 (post-event) topographic maps area constructed using photogrammetric methods (yellow contour lines). Areas of net accumulation are shown in red. Areas of net depletion are shown in blue. The gray shaded area near the toe of the deposit was modified by emergency grading activities in the immediate aftermath of the disaster.

(3) The elongated shape and U-shaped profile of the lower depletion zone mimics the shape of glacially-carved valleys, consistent with the theory that debris flows should exhibit erosional behavior analogous to that produced by glaciers [Johnson, 1970].

Scouring added approximately $5,411 \, \text{yd}^3$ ($4,140 \, \text{m}^3$) of material to the debris flow, corresponding to about 13 percent of the total depletion volume (Table 8.2). Comparison of the total depletion and accumulation volumes indicates a deficit of approximately $5,541 \, \text{yd}^3$ ($4,240 \, \text{m}^3$) in the accumulation figure. This difference appears to correspond to debris removed from Santa Barbara Avenue during the initial emergency response. The average thickness of debris in the accumulation zone is estimated at 6 ft (1.8 m), though the thickness exhibited significant local variation. The thickest accumulations of debris, locally exceeding 15 ft (4.5 m), occurred near the toe of the main lobe (Figure 8.17).

As indicated on Table 8.2, the maximum (horizontal) length of the 2005 La Conchita landslide was 1,335 ft (407 m) between the crown of the headscarp and the toe of the main lobe, corresponding with an elevation drop of 500 ft (152 m). The corresponding average travel path slope or "fahrböschung" [Heim, 1932; Hsü, 1975] was $152/407 = 0.37$ or $\tan(20.5°)$. The latter inclination (20.5°) represents the angle between the toe of the debris flow and the crown of the headscarp. By comparison, the angle of repose of loose rock typically varies between about 32° and 45°. Hence, the debris flow traveled much farther from the mountain front than would be anticipated from a "normal" dry rock landslide and is one likely reason for the high casualty and damage figures resulting from the event. Such "long runout" behavior is characteristic of both wet and dry landslides with volumes exceeding ~ 1.3 million yd^3 ($\sim 106 \, \text{m}^3$), and is pronounced in landslides arising from weak or highly fragmented source materials [Shaller, 1991b]. No scientific consensus yet exists as to the origin of this behavior [Shaller and Shaller, 1996].

8.6.7   *Velocity*

Based on a review of the video coverage of the event, the debris flow locally appears to have been moving at a velocity of around 6 m/s (14 mi/hr) where its western margin was filmed just downslope from the Ranch Road. Likely, the central portion of the flow was moving somewhat faster [~ 10 m/s (~ 22 mi/hr)]. Upon entering the community the velocity was substantially lower, probably 5 m/s (11 mi/hr) or less, based on eyewitness accounts of some residents outrunning the advancing flow.

As an independent check on these estimates, the velocity was estimated using a method set out by Prochaska *et al.* [2008]. This method involves back-calculation of debris flow velocity using superelevation. In this method, fluid pressure is equated to centrifugal force using the forced vortex equation [McClung, 2001]:

$$v = \sqrt{\frac{Rcg}{k}\frac{\Delta h}{b}} \qquad (8.5)$$

where:

v mean flow velocity (m/s)
Rc the channel's radius of curvature (m)
g acceleration of gravity (m/s²)
Δh superelevation height (m)
k correction factor for viscosity and vertical sorting, and
b the flow width (m)

The velocity was estimated for a location near the toe of slope (profile a-a' on Figure 8.18). Key input values were $Rc = 107\,\text{m}$ (351 ft) (see Figure 8.18), $g = 9.8\,\text{m/s}^2$ (32 ft/s²), $\Delta h = 3.2\,\text{m}$ (11 ft), $b = 38\,\text{m}$ (125 ft) and $k = 1$. These input values yield a mean flow velocity, v, of approximately 9 m/s (20 mi/hr), generally consistent with the velocities estimated from the video coverage.

Notably, similar conditions should have prevailed at profile location b-b' (*i.e.*, $\Delta h \sim 3\,\text{m}$ or $\sim 10\,\text{ft}$), though only $\sim 1\,\text{m}$ ($\sim 3\,\text{ft}$) of elevation difference occurs between the paired levees at this location. This finding appears to support the conclusion that debris was shed from the outer (western) edge of the debris flow as it rounded the curve near profile b-b' because the swale was not deep enough to contain the entire flow.

8.7   Previous Studies of Debris Flow Behavior

Debris flows have been described using a variety of mechanical models. The earliest efforts described them as Newtonian fluids [Sharp and Nobles, 1953]. Johnson [1970], however, recognized that debris flows differ fundamentally from Newtonian fluids, in their ability to freight massive debris, to form U-shaped channels, to come to a rest on a slope, to form paired levees and to exhibit snout-shaped distal margins. Some elements of debris flow behavior may be ascribed to a density that is more soil- than water-like, but other observations indicate that the debris itself must exhibit a finite strength. Johnson [1970] characterized debris flows as plastics, materials that behave as rigid bodies under low shear stresses but flow when their shear strength

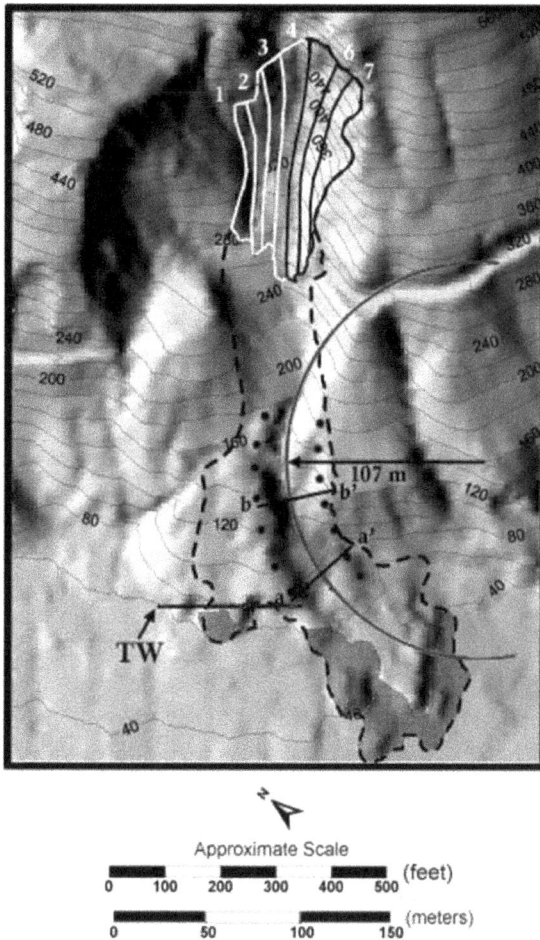

Figure 8.18   Shaded relief map of the 2005 La Conchita landslide with modeling anno-
tation. Black dots show approximate locations of main lobe lateral levees; 107 m radius
circle indicates radius of curvature of levees. Transverse sections a-a' and b-b' show locations of profiles used for superposition calculations. Numbered area at top of page shows
subdivision of headscarp area used for input hydrograph calculations. TW indicates location of temporary wall. See text for discussion.

is exceeded. The minimum shear stress that must be applied to a plastic
material to induce flow is called the yield stress. Examples of materials that
exhibit plastic-type behavior include toothpaste and wet concrete.

Another basic observation of debris flows is that they typically move
more slowly than water floods, and so must exhibit a finite viscosity that is

typically greater than that of water. The simplest mathematical description of such a material is the Bingham plastic, for which a linear relationship exists between the applied shear stress and the shearing rate when the applied shear stress exceeds the yield strength [Davies, 1997].

The rheological modeling of debris flows is very complex. As stated by Davies [1997], "If a debris flow could be modeled as a single, homogeneous material, prediction of its behavior using simulation models would be very much simplified." As further described by Davies [1997], debris flows are strongly influenced by sediment concentration, hysteresis effects, the necessity to maintain a minimum shear rate, complexities introduced by large grains, and complications resulting from a material that exhibits both frictional and fluid behavior.

8.8 FLO-2D Numerical Modeling

8.8.1 *Introduction*

The numerical analyses performed for this investigation utilized the FLO-2D commercial software package, originally developed in 1988 to conduct a Federal Emergency Management Agency (FEMA) flood insurance study of an urbanized alluvial fan in Colorado [FLO-2D Users Manual — O'Brien *et al.*, 2007]. FLO-2D has the capability to model many of the complex behaviors exhibited by debris flows, and has been used to model debris flow hazards elsewhere [*c.f.*, O'Brien *et al.*, 2007], though to the best of our knowledge the La Conchita event is the largest debris flow to have been so modeled. Furthermore, the 5 ft (1.52 m) square grid is the finest resolution that has been used to model a large debris flow, thus possibly making this the largest FLO-2D debris flow simulation ever implemented.

8.8.2 *FLO-2D modeling of debris flows*

The FLO-2D software package allows users to predict the pathways and depths taken by overland flood events. FLO-2D mathematically tracks a flood hydrograph from a series of inflow grids over a system of square grid elements that describe the elevation and roughness of the modeled terrain. The program contains subroutines for modeling hyper-concentrated sediment flows, including mudflows and debris flows, and incorporates the effects of dilution, flow cessation and flow remobilization. FLO-2D implements the Diffusive Hydrodynamic Model (DHM) of Hromadka and Yen [1987], which applies a simple numerical approach with a finite difference scheme that allows the calculation, at each time step, of the attributes of individual grid elements (such as flow depth and velocity).

FLO-2D predicts the behavior of mudflows and debris flows by calculating the shear stress as a summation of five components:

(1) the yield stress resulting from cohesion between fine grains of sediment, τ_c;
(2) the Mohr-Coulomb shear stress, τ_{mc};
(3) the viscous shear stress, which accounts for the fluid-particle viscosity, τ_v;
(4) the turbulent shear stress, τ_t;
(5) and the dispersive shear stress, τ_d, which accounts for particle collisions.

These components, written in the form of shear rates, give a quadratic rheological model that is calculated as a function of sediment concentration [Julien and O'Brien, 1997]:

$$\tau = \tau_y + \eta \left(\frac{dv}{dy}\right) + C \left(\frac{dv}{dy}\right)^2 \tag{8.6}$$

where $\tau_y = \tau_c + \tau_{mc}, C = \rho_m l^2 + f_i(\rho_s, C_v)d_s^2$, and η is the dynamic viscosity. Where ρ_m is the density of the mixture, l is the Prandtl mixing length, ρ_s is the density of the sediment, C_v is the volumetric sediment concentration and d_s is the effective sediment size. More details of the FLO-2D program for debris flow modeling can be found in Julien and O'Brien [1997].

Key steps in the construction of a FLO-2D debris flow simulation include: creation of the model space grid using digital terrain data and supplementary elevation data for engineered structures such as channels and walls; creation of the inflow hydrograph and corresponding sediment concentration; input of surface roughness (Manning's coefficient); and estimation of the weight, yield strength and dynamic viscosity of the debris. The FLO-2D Users Manual [O'Brien *et al.*, 2007] provides a library of yield strength and dynamic viscosity parameters that can be used if these values cannot be independently established.

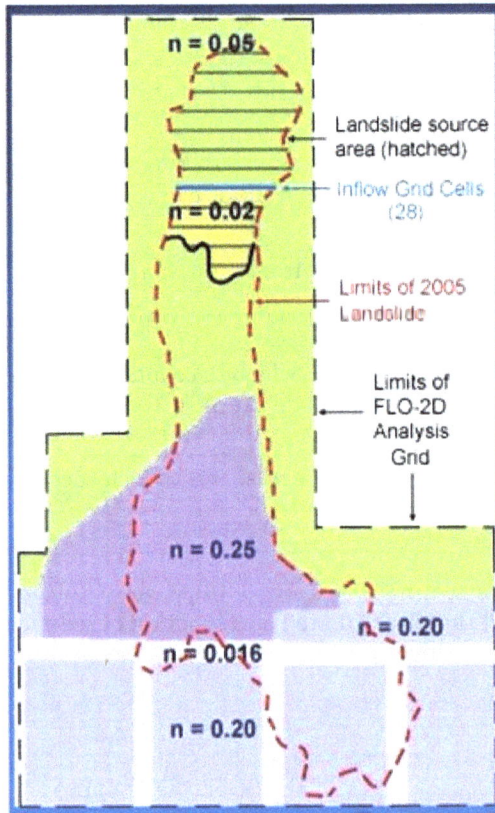

Figure 8.19 Diagram showing limits of FLO-2D simulation grid, the location of the 28 inflow grid cells, and the areal distribution of surface roughness values used in the simulation relative to the limits of the 2005 deposit.

8.8.3   *Input parameters*

8.8.3.1   *FLO-2D grid*

The initial step in the analysis was preparation of the FLO-2D grid, the model surface over which the flow was to be routed. A grid consisting of 25,615 1.52 m (5 ft) square cells was developed that covered the source and potential runout areas (see Figure 8.19). The small unit cell size was selected to provide the level of detail required to resolve the key questions at issue in the investigation. Topographic input for the grid was developed using the minimum topography resulting from the overlap of the pre- (2002) and post-event (2006) topographic maps. The resulting surface was essentially the 2002 map minus the areas of depletion shown on Figure 8.17. This model surface was chosen to represent the conditions experienced by the majority of the debris flow over the course of a simulation. It was reasoned that the wet or saturated material in the lower depletion zone was probably eroded out in the early stages of the debris flow to firmer underlying material. Hence, the analysis was performed using this hybrid topographic surface. Once the topography was rendered onto the grid, the height of the temporary wall was entered manually based on as-built drawings of the structure.

8.8.3.2   *Inflow hydrograph*

The inflow hydrograph represents one of the key pieces of input data for the analysis. For modeling of flood events, the inflow hydrograph represents the amount of water passing into the model space through the inflow grids as a function of time. In the modeling of debris flows, an additional factor, the sediment concentration, is introduced. The sediment concentration, like the volume of water, may be varied over the duration of the inflow hydrograph.

Four key elements were required to generate the inflow hydrograph for the La Conchita event: (1) the volume of water and sediment; (2) the duration of the hydrograph; (3) the variation in the concentration of solids vs. water as a function of time from initiation; and (4) the initial lateral distribution of debris emerging from the headscarp area.

The volume of debris input into the simulations was taken as the full volume of both the upper and lower depletion zones. It was reasoned that the entire volume should be incorporated into the model because it represented the ultimate mass of debris moved in the event. This entire volume was added into the inflow hydrograph and input into the model along a series of 28 inflow grid cells distributed across the upper zone of depletion. The input grid cells were placed at a central location within the upper depletion zone to average out the effects created by the distributed nature of the debris in the source landslide.

Evidence from the television videotape indicates that the debris flow was moving as fast as $\sim 10\,\mathrm{m/s}$ (22 mi/hr) near the toe of the upper depletion area. Based on the surface topography in this area, it was assumed that the initial landslide accelerated from rest to $\sim 10\,\mathrm{m/s}$ (22 mi/hr) over a slope distance of approximately 37 to 74 m (121 to 243 ft). These considerations imply that the acceleration phase took between 8 and 16 seconds. For these simulations, a 14-second duration was adopted for the inflow hydrograph.

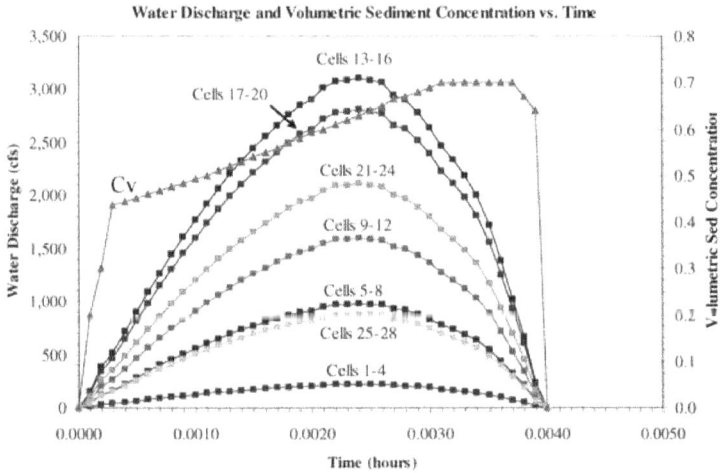

Figure 8.20   Volume sediment concentration (Cv) and inflow hydrographs used for the FLO-2D simulations.

The dense, coherent appearance of the moving debris captured on video implies that the debris flow had a very high sediment concentration throughout the imaged part of the event. Based on suggested parameters contained in the FLO-2D Users Manual [O'Brien *et al.*, 2007], the sediment concentration in the input hydrograph was therefore rapidly increased from zero (the starting point required by the model) to a value of 0.5, then ramped up more slowly to a peak value of 0.7 over the 14-second duration of the hydrograph. Figure 8.20 shows the temporal variation in the volume concentration of solids utilized in the input hydrograph.

The upper zone of depletion exhibits a spoon-like shape, indicating that the debris did not exhibit a uniform lateral distribution at the initiation of the event. This geometric effect was modeled by subdividing the source area into seven parallel chutes, as shown in Figure 8.18, with each chute feeding its prorated volume of debris into four inflow grid cells. Table 8.3 shows the percentage of the total volume of debris assigned to each chute and Figure 8.20 shows the shape of the actual inflow hydrographs associated with each input grid cell (seven groups of four cells each).

8.8.3.3   *Unit weight*

Samples of debris collected from test pits in the western lobe of the debris flow were tested to assess the moisture content and dry density of this material. These results were used to estimate a saturated density of

Table 8.3   Distribution of mass from source area chutes.

1	2	3	4	5	6	7
2%	8%	14%	26.5%	24%	18%	7.5%

1,762 kg/m^3 (110 lb/ft^3) for the debris at the time of the event. This value was used in all subsequent calculations.

8.8.3.4   *Yield strength*

An initial estimate of the plastic yield strength of the debris, a necessary input to the FLO-2D simulations, was calculated using parameters derived from the pre- and post-event topographic maps of the area by means of the following relationship [Johnson, 1970]:

$$\tau = Tc\gamma d \sin \delta \qquad (8.7)$$

where:

   τ yield strength (Pa)
   Tc critical thickness (m)
   γ^d unit weight of debris (kg/m^3)
   δ slope angle (degrees)

The yield strength of the debris flow was calculated for each 1.52 m (5 foot) grid in the zone of accumulation using the calculated thickness of the debris (see Figure 8.17), the pre-event (2002) slope angle, and a unit weight of 1,762 kg/m^3 (110 lb/ft^3). Because this approach produced unrealistically high strength values for debris captured in narrow channels, the median (rather than mean) value of the calculated yield strength, 5,257 Pa (110 lb/ft^2), was adopted. This strength was taken to be equivalent to the yield strength of the debris at a volume concentration of solids of 0.59, considered representative of the average debris at the time of emplacement. The adopted yield strength parameters used in the FLO-2D analysis are reported in Table 8.4. Figure 8.21 shows a comparison of the adopted yield strength values with published values from the FLO-2D Users Manual [O'Brien *et al.*, 2007].

8.8.3.5   *Dynamic viscosity*

Because no independent means was available to estimate the dynamic viscosity of the debris from the final deposit, an intermediate value was

Table 8.4   Variables adopted for FLO-2D analysis.

Variable	Yield Stress $\tau = \alpha \cdot e^{\beta C_v}$	Dynamic viscosity $\eta = \alpha \cdot e^{\beta C_v}$
α	1.75	0.000602
β	17.475	22.5

adopted from values contained in the FLO-2D Users Manual [O'Brien *et al.*, 2007]. The adopted dynamic viscosity parameters used in the FLO-2D analysis are reported in Table 8.4. Figure 8.21 shows a comparison of the adopted dynamic viscosity values with other published values from the FLO-2D Users Manual [O'Brien *et al.*, 2007].

8.8.3.6 *Surface roughness*

Review of the 2002 aerial photography (Figure 8.13) indicated that a variety of surface roughness values were appropriate for input to the FLO-2D simulations. As shown on Figure 8.19, roughness values ranging from 0.016 to 0.25 were used in the analysis. Low roughness values were used for streets in the development (0.016) and to model the portion of the headscarp area located downslope from the inflow grid cells (0.02). High roughness values were added for the residential area (0.2) and for the area of dense chaparral located just upslope (0.25). Intermediate values (0.05) were used for the remainder of the undeveloped slope area, which supported a moderate vegetation cover.

8.8.3.7 *Modeling of temporary wall*

Three cases were considered to model the potential effects of the temporary wall on the debris flow. The first was to run the simulation with the wall as it existed prior to the failure (termed "With Wall"). Because there is no mechanism within FLO-2D to simulate the destruction of a structure, this simulation case assumes that the wall exhibits infinite strength. In the second case, the FLO-2D grid was manually modified to remove the topographic effect of the wall in areas where it was observed to have been overtopped and/or breached by the debris flow (termed "Breached Wall"). Elsewhere the wall continued to be treated as if it exhibited infinite strength. In the third case, the FLO-2D grid was modified to remove the temporary wall entirely and restore the topography present before the construction of the wall (termed "No Wall"). A 1996 topographic map was

YIELD STRESS & DYNAMIC VISCOSITY

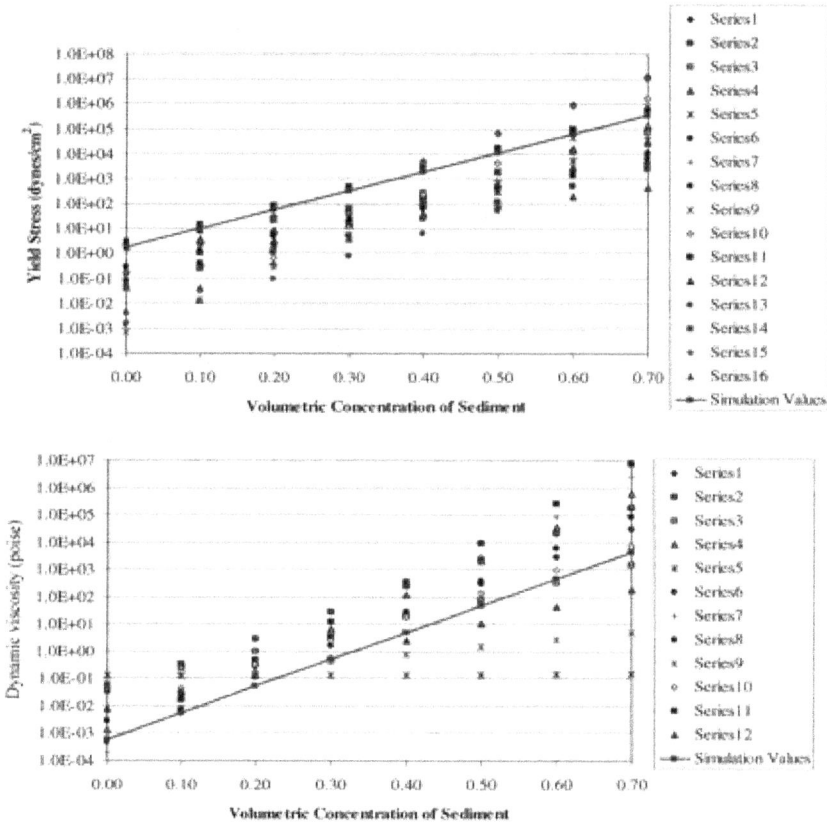

Figure 8.21   Graphs showing adopted yield strength (top) and dynamic viscosity (bottom) relationships adopted for the FLO-2D simulations plotted against literature values reported in the FLO-2D Users Manual [O'Brien *et al.*, 2007].

utilized to establish the topography of the area to the earlier, pre-wall condition.

8.8.4   *Model results*

8.8.4.1   *General observations*

Figure 8.22 presents the results of the baseline FLO-2D simulations for the three cases (sensitivity analyses are discussed separately). The three simulations consistently predict the formation of a large main (eastern)

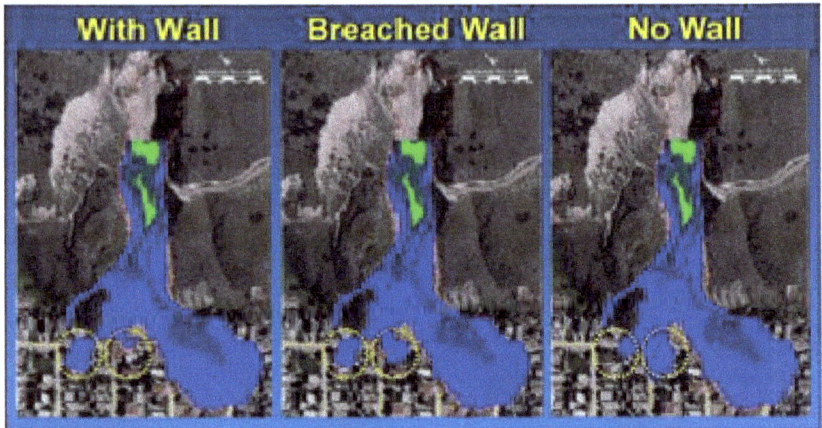

Figure 8.22 FLO-2D simulation for the "With Wall," "Breached Wall," and "No Wall" conditions. The limits of the actual deposit are shown with red dashes. As shown above, differences between the three starting conditions have very little effect on the predicted behavior of the eastern (main) lobe, but result in significant differences in the predicted behavior of the western (minor) lobe. The color shading reflects the predicted thickness of the debris, ranging from blue (minimum) to light green (maximum).

lobe, a smaller, more irregular minor (western) lobe, and a finger of material filling the pre-existing drainage channel in the mid-slope area.

8.8.4.2 *Main (Eastern) lobe*

The simulations consistently show that the modeled wall condition exercised a minimal effect upon the distribution of the main lobe of debris. The visible effects of the wall were limited to the extreme western margin of the main lobe near Vista del Rincon Avenue. Otherwise, the three baseline FLO-2D simulations predict very similar behavior for this lobe of the debris.

Notably, all three simulations consistently predict debris in the main lobe spreading one row of houses farther to the east than actually occurred. At least four factors likely contributed to the overprediction of debris inundation in this area:

(1) Incorporation of the full depletion volume (including scoured material) in the inflow hydrograph, thereby raising the average potential energy of the debris in the simulation relative to the actual conditions;

(2) Using the scoured topography in the FLO-2D input grid, resulting in a clearer and lower travel path for the debris than under actual conditions, resulting in less material overflowing into the minor lobe;

(3) Conceptual difficulties in modeling the interaction of the debris flow with houses and other large obstacles in the community; and,

(4) Simulation of the final stages of movement.

With respect to the final point, all the FLO-2D simulations predicted very slow movement of the debris upon reaching the vicinity of Santa Barbara Avenue. Such slow movement may be unrealistic because the elevated pore pressures required to keep the debris in a fluid state are maintained by the agitation of the moving debris, which is in turn a function of its velocity. As the velocity and agitation of the debris diminishes, the pore pressures will diminish in kind, ultimately falling below that needed to keep the debris in a fluid condition. A "critical velocity" necessary to maintain fluidity was not, however, determined in this analysis.

8.8.4.3   *Minor (Western) lobe*

The most significant differences among the FLO-2D simulations and between the simulations and the actual behavior of the debris flow occur in the area of the western lobe. The most dramatic differences are evident at the western end of the temporary wall (left-hand circled area on Figure 8.22), where the "With Wall" and "Breached Wall" simulations both predict a lobe of debris spilling out from behind the wall that did not occur in the actual event. The FLO-2D prediction appears to result from the simulated remobilization of material that accumulates behind the wall. In the actual event, the debris that accumulated behind the wall arrived in pulses and came to rest. Evidently, this allowed the elevated pore pressures in the debris to dissipate, increasing its strength and viscosity such that it did not remobilize as predicted in the simulations.

The FLO-2D simulations also diverge from each other and from the actual behavior of the debris flow near the center of the temporary wall (right-hand circled area on Figure 8.22). The small lobe associated with the "With Wall" simulation is consistent with the anticipated behavior of the debris where the wall was present and indestructible; only a small amount of debris spills over the wall at a local low spot.

In the "Breached Wall" simulation, debris flows through the breach, but extends somewhat farther than it did in the actual event. Again, this can probably be ascribed to the pulse-like arrival of debris to the wall in the actual event and the interaction between the wall and the debris. Although the wall was breached by the impacting debris, the material was slowed sufficiently to modify its rheology and diminish its mobility.

In the "No Wall" simulation, the debris in the minor lobe is predicted to travel considerably farther than occurred in the other simulations and in the true event. This simulated behavior appears to reasonably approximate the distribution of debris that would have occurred had the event occurred prior to 2000 (*i.e.*, absent the temporary wall and the removal of debris from Vista del Rincon Avenue). This result suggests that the installation of the wall and the removal of debris from the roadway, though not intended to mitigate landslide hazards, protected one or two houses from the impact of debris that would likely have occurred in the absence of this work.

Finally, the simulations all predict the occurrence of a small clear area between the main and minor lobes that was in fact mantled by the debris flow. This difference between actual and predicted behavior was probably related to the incorporation of the lower depletion zone into the FLO-2D simulation grid. In the actual event, early arriving material in the main lobe was at a higher elevation and, as a result, could more readily spill over into this area due to superelevation as the debris flow curved to the left.

8.8.4.4 *Channel fill*

All of the FLO-2D simulations predict a finger of debris extending down the incised channel in the mid-slope area. Review of the simulation output indicates that this is water rich material mobilized in the early stages of the landslide. Later-arriving material exhibiting a higher sediment concentration subsequently clogs the steep-sided channel, resulting in avulsion of debris from the channel. This behavior is generally consistent with field observations of the channel downstream from the portion choked by the debris flow.

8.8.4.5 *Sensitivity analyses*

A sensitivity analysis was conducted for each of the FLO-2D simulations, in which the yield strength and dynamic viscosity were independently varied by ±20 percent. These variations had a minor effect on the behavior of the debris in the simulations, with changes to the yield strength having the larger effect. A 20 percent reduction in the yield strength resulted in a maximum of 12 m (40 ft) of additional runout of the main lobe compared to the baseline case, whereas a 20 percent increase in strength reduced the runout by up to about 9 m (30 ft). By comparison, variation of the dynamic viscosity by ±20 percent resulted in a change of about ±3 m (±10 ft) in the runout distance.

8.9 Summary

Intense rainfall coupled with unfavorable geologic conditions triggered the 2005 La Conchita landslide, which transformed into a giant debris flow containing over $30{,}000\,\mathrm{m}^3$ $(40{,}000\,\mathrm{yd}^3)$ of wet soil and rock. The main lobe of the debris flow, containing 90% of the mobilized debris, entered a residential neighborhood, killing 10 persons and damaging or destroying 36 residences. The debris flow was modeled using FLO-2D software to assess the mechanics of the debris flow and the role, if any, played by a temporary wall in altering the path taken by the debris flow as it traversed the community. This investigation included construction of the largest, most detailed FLO-2D model attempted for analysis of a debris flow. The model simulations offered a good approximation of the actual behavior of the debris flow. Based on the modeling results, the destructiveness of the event can be attributed to the large volume of debris mobilized, the geometry of the flow path, and the yield strength and viscosity of the flowing debris, in relative order of importance. Differences between the model and actual behavior principally resulted from the fixed character of the terrain over which the flow was routed, the behavior of the debris at low velocities and the remobilization of stopped material. The different model runs provided useful insights into the behavior of the debris flow as it descended the hillside, as well as its interaction with the temporary retaining wall.

References

Davies, T.R.H. (1997). "Using hydroscience and hydrotechnical engineering to reduce debris flow hazards." In Debris-Flow Hazards Mitigation: Mechanics, Prediction and Assessment, C.-J. Chen, (ed.), ASCE, New York, N.Y., 787–810.

Gurrola, L.D. (2005). "Recent landslides in La Conchita, California belong to a much larger prehistoric slide, report geologists." U. C. Santa Barbara, press release (October 19).

Harp, E.L., Jibson, R.W., Savage, W.Z., Highland, L.W., Larson, R.A. and Tan, S.S. (1995). "Landslides triggered by January and March 1995 storms in southern California." Landslide News, 9, 15–18.

Heim, A. (1932). "Bergsturz und Menschenleben." Fretz and Wasmuth A.G., Zurich, 218 p. (English translation by N.A. Skermer, BiTech Publishers, Vancouver, BC).

Hemphill, J.J. (2001). "Assessing landslide hazard over a 130-year period for La Conchita, California." Proc. Assoc. Pacific Coast Geogr. Ann. Mtg, Santa Barbara, CA, September 12–15, 2001.

Huftile, G.J., Lindvall, S.C., Anderson, A., Gurrola, L.D. and Tucker, M.A. (1997). "Paleoseismic investigation of the Red Mountain fault: Analysis and trenching of the Punta Gorda terrace." On-line document accessible at: http://www.scec.org/research/97research/97huftilelindvalletal.html

Hulme, G. (1974). "The interpretation of lava flow morphology." *Geophys. J. R. Astr. Soc.*, 39, 361–383.

Hromadka, T.V. and Yen, C.C. (1987). "Diffusive hydrodynamic model." U.S. Geological Survey Water Resources Investigations Report 87–4137.

Hsü, K.J. (1975). "Catastrophic debris streams (sturzstroms) generated by rock-falls." *Bull. Geol. Soc. Am.*, 86, 129–140.

Johnson, A.M. (1970). "Physical processes in geology." Freeman, Cooper and Co., San Francisco, Calif., 433–459.

Julien, P.Y. and O'Brien, J.S. (1997). "On the importance of mud and debris flow rheology in structural design." In Debris-Flow Hazards Mitigation: Mechanics, Prediction and Assessment, C.-I. Chen, (ed.), ASCE, New York, N.Y., 350–359.

McClung, D.M. (2001). Superelevation of flowing avalanches around curved channel bends," *J. Geophys Res.*, 106(16), 16,489–16,498.

National Climatic Data Center (2005). "2004/2005 winter storms: California and the southwest U.S." www.mhsweather.org/images/Meteorology _2005_Jan_Feb_ So_Cal_Rain_Events.doc

O'Brien, J., Jorgensen, G. and Garcia, R. (2007). FLO-2D Software and Users Manual, Nutrioso, AZ.

O'Tousa, J. (1995). "La Conchita landslide, Ventura County, California." AEG News, 38/4, 22–24.

Prochaska, A.B., Santi, P.M., Higgins, J.D. and Cannon, S.H. (2008). "A study of methods to estimate debris flow velocity." *Landslides*, 5, 431–444.

Rogers, J.D., Watkins, C.M., Rock, F., Kane, W.F., Owen, J. and Bell, M. (2007). "The 1928 St. Francis Dam Failure and the 1995/2005 La Conchita landslide: The Emergence of engineering geology and its continuing role in protecting society." Guidebook, Assoc. of Env. & Eng. Geol., 50th Anniv. Ann. Mtg., Los Angeles, CA, September 24–29, 2007, 52 p.

Schiek, C.G. and Hurtado, J.M. (2005). "Analysis of the 1995 and 2005 La Conchita, CA Landslides using Aerial Photographs and ASTER Satellite data." AGU Fall Meeting, abstract #G11A-1179.

Shaller, P.J. (1991a). "Analysis of a large moist landslide, Lost River Range, Idaho, USA." *Can. Geotech J.*, 28, 584–600.

Shaller, P.J. (1991b). "Analysis and implications of large martian and terrestrial landslides." Ph.D. dissertation, California Institute of Technology, 586 p.

Shaller, P.J. and Shaller A. (1996). "Review of proposed mechanisms of Sturzstroms (long-runout landslides)." In: Sturzstroms and Detachment Faults, Anza-Borrego State Park, California, P.L. Abbot and D.C. Seymour, (eds.), South Coast Geological Society, Annual Field Trip Guidebook No. 24, pp. 185–202.

Sharp, R.P. and Nobles, L.H. (1953). "Mudflow of 1941 at Wrightwood, southern California." *Bull. Geol. Soc. Am.*, 64, 547–560.

U.S. Department of Agriculture (2008). On-line Web Soil Survey, http:// web-soilsurvey.nrcs.usda.gov/app/WebSoilSurvey.aspx

U.S. Geological Survey (2008). USGS On-line Video and Image Gallery, glacier images, http://gallery.usgs.gov/tags/glacier/list/eiy4Cpo00V_3/1

Tiger Wash, Western Maricopa County, Arizona, USA

Jonathan E. Fuller

JE Fuller/Hydrology & Geomorphology, Inc.
8400 S Kyrene Rd., Suite 201, Tempe, Arizona 85284
jon@jefuller.com

The Tiger Wash alluvial fan has served as an excellent laboratory for studying alluvial fan flood processes. Information collected at the site before and after a very large flood in September 1997 flood sheds light on the nature of alluvial fan flooding and avulsive channel change on fluvially-dominated fans in semi-arid climates. The information collected from the Tiger Wash fan also helps clarify broader questions such as: (1) What is an alluvial fan? (2) What are the essential characteristics of alluvial fan flooding? The studies conducted on the Tiger Wash alluvial fan also provide insight on very practical matters such as: (1) identifying alluvial fan boundaries, (2) the frequency of channel avulsions on the fan surface, (3) the processes by which channel avulsions occur, (4) the importance of runoff infiltration on the fan surface, and (5) the accuracy of floodplain delineation techniques which reflect actual flood hazards on alluvial fans. Most importantly, the research performed on the Tiger Wash alluvial fan highlights the need for close interaction between engineers, geologists and floodplain managers when assessing alluvial fan flood hazards.

8.10   Site Description

8.10.1   *Watershed*

The Tiger Wash alluvial fan is located in far western Maricopa County, Arizona (Figure 8.23). The $250\,\mathrm{km}^2$ ($96.4\,\mathrm{mi}^2$) watershed drains to the fan's topographic apex where it diverges into two branches informally named the

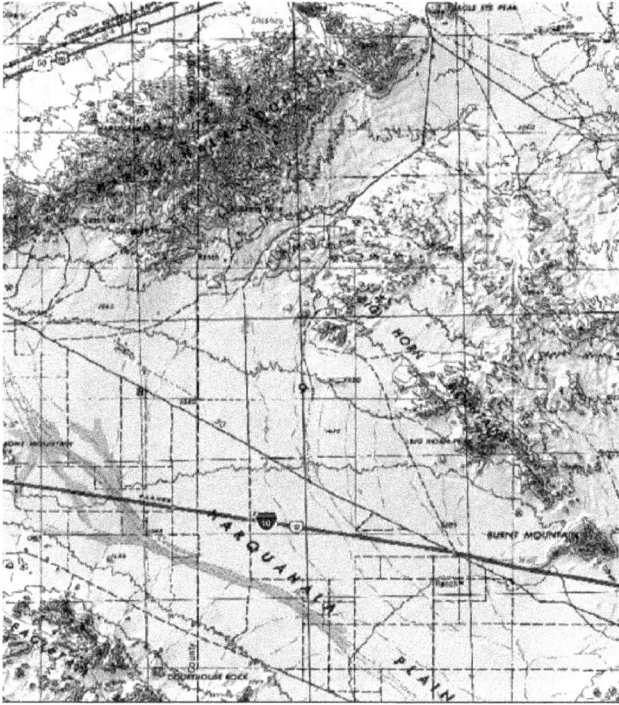

Figure 8.23 Tiger Wash alluvial fan and watershed.

East Branch and West Branch. The East Branch flows nearly due south, while the West Branch flows to the southwest into the alluvial fan study area (Figure 8.24).

The climate in the study area is semi-arid, with annual precipitation averaging less than 254 mm (10 in) near the fan apex, with increased rainfall depths at higher elevations in the watershed. Precipitation is typically divided equally between summer and winter seasons. Summer storms are associated with warm, moist tropical air masses from the Gulfs of Mexico and California that produce intense localized thunderstorms. Winter precipitation usually originates over the Pacific Ocean and produces light to moderate rain over large geographic areas. A third major precipitation source is dissipating tropical storm remnants, which generally occur in late September and October, and generate moderate to intense rainfall that can last for many hours.

Figure 8.24   Aerial photograph of Tiger Wash with gauge locations.

8.10.2   *Geologic setting*

Tiger Wash is located within the Basin and Range Physiographic Province, a region characterized by northwest trending mountain ranges and intervening alluvial valleys. The Tiger Wash watershed has been tectonically quiescent for at least 6 million years [Shafiqualla *et al*, 1980] (Figure 8.25). Climate change has been the primary cause of any sediment deposition on the alluvial fan during the Pleistocene [Demsey, 1989]. Tiger Wash drains portions of Harquahala and Big Horn Mountains. The Harquahala Mountains consist of a granitic basement overlain by quartzite and marine sedimentary rocks. The Big Horns are composed primarily of volcanic sedimentary rock. A number of basaltic inselbergs penetrate the alluvial fill of the Tiger Wash piedmont, suggesting a relatively shallow depth of alluvium and potentially a pediment surface upslope from the alluvial fan apex. Aside from the inselbergs, no evidence of bedrock outcrop was observed on the surface of the alluvial fan.

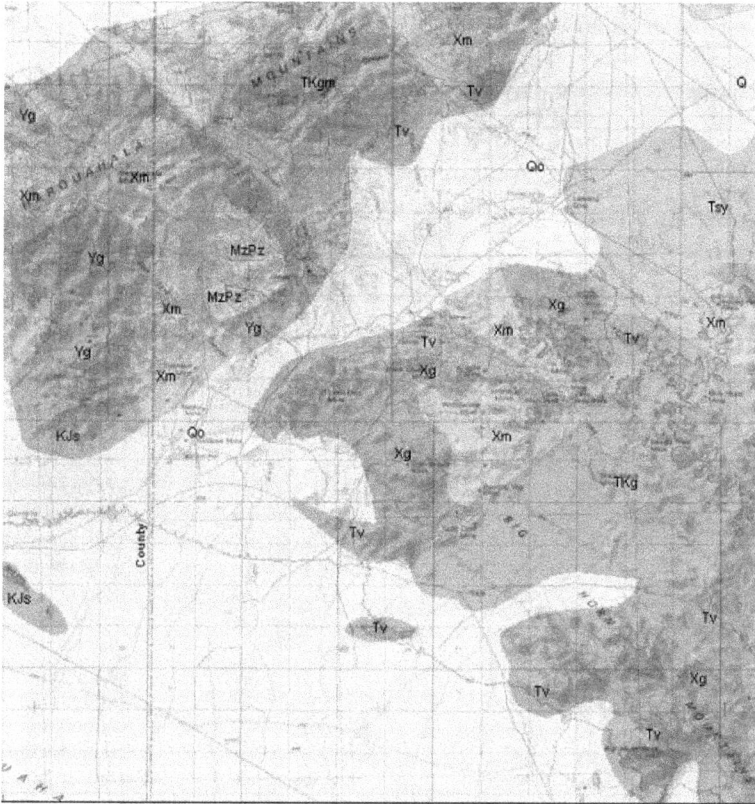

Figure 8.25   Geologic map including surficial geologic mapping.

8.10.3   *Surficial geology*

The surficial geology of the Tiger Wash piedmont was mapped in detail by Pearthree *et al.* [2004] (Table 8.5; Figure 8.26). Holocene mapping units were divided into: (1) Q_{yc}, the active fluvial system; (2) Q_{y2}, overbank and sheetflood areas; and, (3) Q_{y1}, older deposits that have been part of the active depositional system recently but which now appear to be isolated from flood inundation. Q_{y2} surfaces are finer grained than Q_{yc} channel deposits, but have very weak soil development, reflecting the periodic deposition of fresh sediment during floods. Q_{y1} surfaces have minimal rock varnish and desert pavement development and are weakly dissected, with variable surface relief. Relict distributary channels ("ghost channels") can be observed on Q_{y1} surfaces, some of which have been integrated into

Table 8.5   Characteristics of surficial geologic units.

Surface map unit	Drainage pattern	Sediment	Soils
Q_{yc}: Modern channels	Single, braided, distributary	Very poorly sorted sand — cobbles	Layered, no soil development
Q_{y2}: Sheet flood and terraces	Poorly defined, discontinuous small	Sand and silt, with local fine gravel deposits	Depositional layering, weak soil development
Q_{y1}: Young fans and terraces	Distributary channels, local swales	Poorly sorted sand — cobbles,	Minor soil development, some carbonate, minimal varnish
Q_l: Relict fans and terraces	Local tributary channels, minor distributary	Poorly sorted cobbles, pebbles	Varnish, desert pavement; moderate soil development, slight reddening, carbonate coatings
Q_m: Old relict fans	Entrenched tributary networks	Very poorly sorted cobbles, pebbles, sand, small boulders	Strong varnish, desert pavements, moderate soil development, reddened, carbonate cementation
Q_{mo}: Very old relict fans	Deep dissected tributary networks	not described	Strong soils, varnish, pavement, carbonate, reddening

the local on-fan tributary drainage networks. The broad area covered by Q_{y1} deposits on the piedmont imply that substantial changes in the loci of flooding and deposition have occurred during the past few thousand years.

Pleistocene-aged surfaces on the piedmont record the longer-term evolution of the Tiger Wash alluvial fan, and include Q_l, Q_m, and Q_{mo} units, from youngest (latest Pleistocene) to oldest (middle Pleistocene). Pleistocene surficial geologic units typically have increasing rock varnish, desert pavements, and soil profile development with age. Q_{mo} surfaces only occur at the margins of the alluvial fan above the apex. Q_l and Q_m surfaces represent relict alluvial fan areas that have been isolated from active deposition for at least 10,000 years. In the upper piedmont, these surfaces

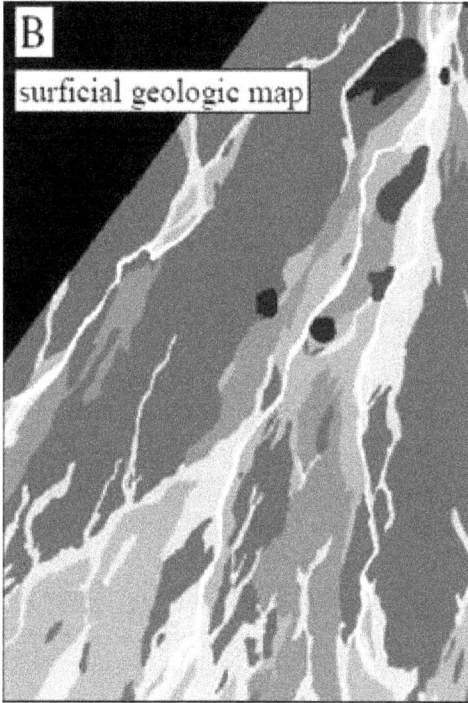

Figure 8.26 Surficial geologic mapping (AZGS).

are substantially higher than adjacent Holocene surfaces. In the middle and lower piedmont, topographic relief between Holocene and Pleistocene surfaces is minimal, and in some situations the Q_1 and Q_m surfaces are actually lower than the adjacent Q_{y2} and Q_{y1} surfaces. Some low-lying Q_m surfaces were partially buried during recent floods. The observations indicate that margins of the active alluvial fan areas evolve over time, and the contacts between young and old surfaces are modified during large floods.

The distribution of surficial units points to areas that are most likely to be subject to alluvial fan flooding. Young Holocene deposits are limited in the northern part of the study area, where Tiger Wash is a tributary drainage system that is topographically confined by bedrock hills and Pleistocene alluvial fan deposits. More extensive Holocene deposits exist along both branches of Tiger Wash in the area of the first major distributary channel split, but very young deposits (units Q_{y2} and Q_{yc}) are restricted to narrow areas along the main channel systems. Q_{y2} deposits are more extensive along the East Branch, which suggests that this branch may have been more important in the past. During the most recent large floods, most of the flow went down the West Branch, and much of the area covered by Q_{y2} deposits on the East Branch was not inundated. The lack of extensive young surfaces near the most upstream distributary split implies that substantial changes in depositional patterns have occurred recently. The moderate topographic relief between active channels and adjacent fan surfaces [up to 3 m (10 ft)] and the existence of relict Pleistocene alluvial fan surfaces within and bounding the distributary system implies that the general configuration of the distributary channel system in this area is reasonably stable.

Downstream of the distributary split into the East and West Branches, the lateral extent of Q_{y2} deposits increases dramatically. In addition, the channel systems on the Q_{y2} surfaces branch and become smaller downstream, and local topographic relief declines to less than about 1 m (3 ft). The areas of extensive Q_{y2} deposits and downstream-branching channel networks are considered to be active alluvial fans within a larger distributary drainage system. Most Q_{y2} surfaces are inundated by sheet flooding in large floods, with potential for significant channel pattern changes. Down-fan Q_{y2} deposits become more extensive and channels become narrower and discontinuous.

8.10.4   *Channel morphology*

At its primary apex, Tiger Wash is a fifth order sand and gravel bedded ephemeral stream with a slightly sinuous braided channel pattern, moderate entrenchment, and a high width/depth ratio. Downstream of the apex, the drainage pattern becomes highly distributary with decreasing channel size and extensive overbank sheet flooding. Individual channels on the fan surface have low sinuosity (1.02–1.05), well-vegetated low banks, and sand-gravel bed material which decreases in size downfan. Individual and cumulative channel capacity varies wildly with distance, but generally decreases in the downfan direction, due to decreasing bank height and channel width. Channel slope does not decrease significantly across the fan surface, although an obvious nick in the profile occurs near the pre-1997 hydrologic apex (Figures 8.27 and 8.28). Hydraulic model estimates of decreased channel capacity in the downfan direction are supported by anecdotal evidence of widespread sheet flooding crossing the Salome Highway near the alluvial fan toe.

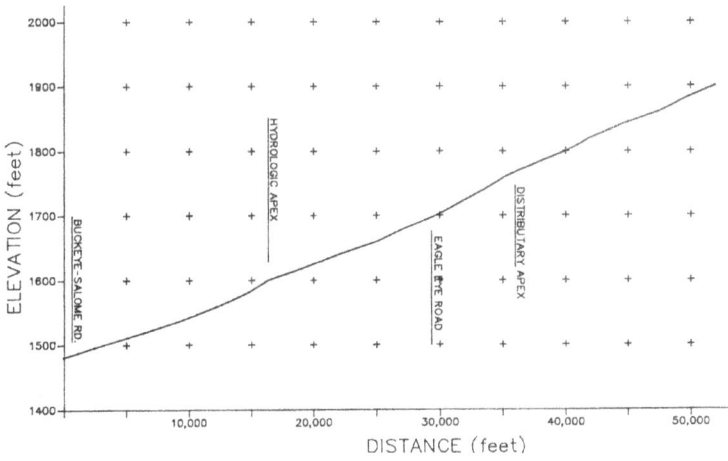

Figure 8.27   Longitudinal profile of the Tiger Wash piedmont.

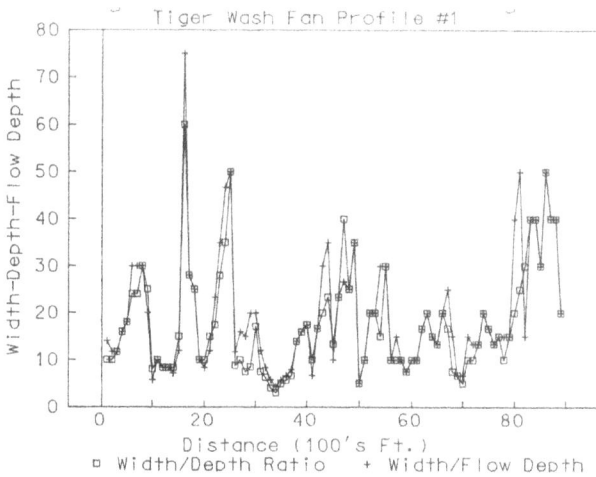

Figure 8.28   Channel geometry variation with distance downfan.

8.10.5   *Outfall*

The Tiger Wash piedmont is truncated by Centennial Wash, a poorly defined axial stream located in the center of the Harquahala Plain. Currently, runoff from Tiger Wash is intercepted by the Central Arizona Project canal levees located upstream of Interstate 10.

8.11 Flood History

Flood information available for Tiger Wash includes a flood-flow-frequency analysis of USGS gauge data, precipitation gauge data, regional regression equation estimates, paleoflood data, and inundation mapping of the 1997 flood. These analyses indicated that the 1997 flood reasonably approximated the 100-year flood.

8.11.1 *Gauge record*

The U.S. Geological Survey (USGS) operates a gauging station (#09517280; Tiger Wash near Aguila, AZ) on Tiger Wash (Table 8.6). The gauge is a manually read crest-stage gauge with a period of record from 1963–1979, 1983, and 1991 to the present [Pope *et al.*, 1998].

In addition to the USGS gauging station, the Flood Control District of Maricopa County (FCDMC) operates three ALERT rain gauges within the watershed. Only one FCDMC gauge has a record longer than 13 years. During the 26 years of its operation, the highest measured 24 hour total rainfall was 96.0 mm (3.78 in) on September 26, 1997. For the same event, the other FCDMC gauges recorded 105.9 mm (4.17 in) and 304.0 mm (11.97 in) of rainfall, with the latter measurement exceeding the highest National Weather Service 24-hour rainfall recorded in Arizona. NOAA Atlas II data indicate a 100-year 24-hour point rainfall value for the Tiger Wash watershed of about 105.4 mm (4.15 in), almost exactly the 24-hour rainfall total at the FCDMC gauge nearest the Tiger Wash alluvial fan apex.

8.11.2 *Peak discharge estimates*

The USGS published two flood flow frequency estimates at the Tiger Wash gauging station, as well an estimate based on a regional regression equation (Table 8.7). The September 1997 flood peak exceeds the USGS 100-year flood frequency estimate, and is less than the USGS regression equation estimate peak, though it lies well within the standard error of the regression equation. The maximum stage of the 1997 flood at the USGS gauging station was about 0.6 m (2 ft) lower than the maximum paleoflood stage indicators identified by CH2M HILL [1992], which corresponded to a maximum paleoflood magnitude of 280 to 370 cubic meters per second (cms) (10,000 to 13,000 cfs). CH2M HILL reported evidence of at least three floods between 170 and 280 cms (6,000 and 10,000 cfs) in the past 100 years prior to 1992.

Table 8.6   Gauged annual peak discharges for Tiger Wash near Aguila, USGS gauge #09517280.

Year	Discharge cms (cfs)	Year	Discharge cms (cfs)
8/16/1963	25.8 (910)	9/25/1976	85.0 (3,000)
10/19/1963	11.3 (400)	8/16/1977	24.6 (870)
8/18/1965	47.6 (1,680)	3/1/1978	39.6 (1,400)
9/13/1966	41.1 (1,450)	12/18/1978	1.7 (60)
8/14/1967	17.6 (620)	9/10/1983	89.8 (3,170)
12/19/1967	12.5 (440)	3/27/1991	1.7 (69)
9/14/1969	12.5 (441)	3/3/1992	25.7 (906)
8/20/1970	128.9 (4,550)	1/8/1993	29.5 (1,040)
8/20/1971	56.6 (2,000)	7/18/1994	103.1 (3,640)
8/20/1972	78.4 (2,770)	2/15/1995	5.2 (185)
10/6/1972	49.6 (1,750)	1/1/1996	0 (0)
8/3/1974	1.3 (45)	9/26/1997	228.5 (8,070)
7/30/1975	2.8 (100)	3/26/1998	19.3 (683)

Table 8.7   Flood frequency estimates for Tiger Wash.

Source	cms (cfs)		
	Q_2	Q_{10}	Q_{100}
USGS WRIR 91-4041	28.6 (1,010)	86.7 (3,060)	196 (6,910)
USGS WRIR 98-4225	27.2 (961)	89.5 (3,160)	208 (7,340)
Thomas, 1997	35.2 (1,242)	128 (4,526)	357 (12,601)

The September 1997 flood was, therefore, about as big as any flood on the alluvial fan in the past 100 years.

Repeat survey data at the gauge [Capesius and Lehman, 2002] suggests about 0.6 meters (2 ft) of bed degradation since 1970, which means that the 1997 flood may have been nearer the magnitude of the largest paleofloods despite its somewhat lesser water surface elevation. Given these data, it can be reasonably assumed that the September 1997 flood approximates the 100-year flood on Tiger Wash.

8.11.3   *September 26, 1997 flood*

The 1997 flood inundated essentially all of the young alluvial fan areas on the Tiger Wash West Branch and caused dramatic changes in the channel system. Several large new channels formed during the flood, and numerous pre-existing channels were enlarged or extended. In the upper part of the West Branch where Tiger Wash is contained in a single large channel confined between Q_{y1} and older surfaces, flow was contained in the channel with minor overbank flow. As channel capacity decreased downfan, flow became unconfined and inundation was widespread in the active alluvial fan areas. Deeper flow occurred in preexisting channels and in overbank areas adjacent to the main channels. Flow in channels transported bedload up to small boulders. Overbank sheet flow transported some gravel, but mainly sand-size and finer particles derived from the main channels, which was deposited as broad sheets and pendant bars downstream of floodplain vegetation. Areas of deep unconfined flow were characterized by alternating small, narrow channels and sand bars, which results in a corrugated topography. In areas of shallow sheetflooding, low sand deposits record flow and deposition around brushy vegetation. At the margins of flow, thin silt layers and flotsam typically cover the surface.

In several cases, flow not only inundated Q_{y2} surfaces, but also formed new channels along on-fan drainages that were small or weakly defined prior to the flood. These channel changes are illustrated dramatically by comparing pre-flood and post-flood photos. The largest new channel formed approximately (Figure 8.29) followed the course of a small channel that existed within an area of fine-grained late Holocene overbank

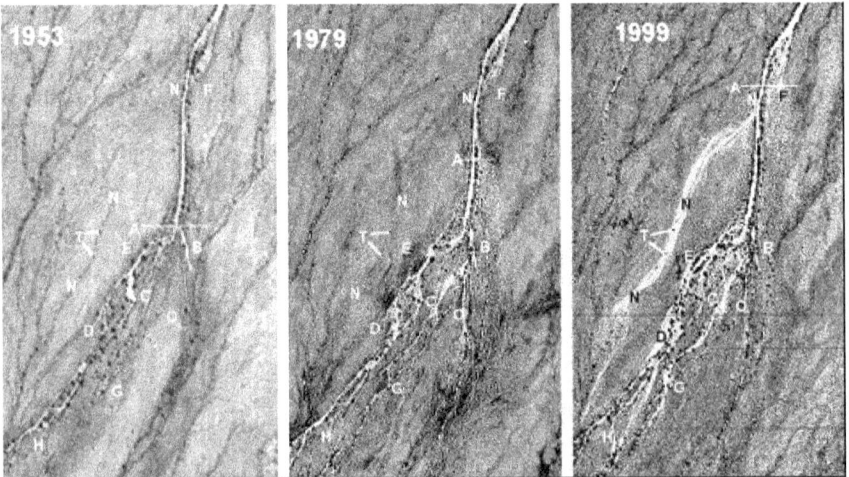

Figure 8.29   Repeated aerial photographs from 1953–1999. Letters identify similar positions in each photograph where dramatic changes have occurred. N = future position of the new avulsive channel. A = positions of the primary influence.

deposits (Q_{y2}). The new channel formed as floodwater inundated the over-bank and downcut into the Holocene deposits. Flow funneled through the new channel, eroding the bed down to Pleistocene-aged soils in places and laterally eroding its Holocene-aged channel banks. This new channel is much wider and somewhat shallower than the pre-existing main channel in that area, but may narrow and deepen with time if larger riparian vegetation becomes established along it.

There are also numerous examples of less dramatic channel changes where flood waters widened or lengthened preexisting channels. These changes occurred in areas of channel flow and deep unconfined flow, flow depth zones where both erosive and depositional settings exist. Changes include channel widening, bank erosion along outside bends, bed scour, and formation of discontinuous well-defined channels. In addition, the apex of the most prominent active alluvial fan on west branch of Tiger Wash has migrated about 500 m (1500 ft) upslope between 1953 and 1999. About half of this migration occurred as a result of the formation of the new western channel in 1997, but it appears from the 1979 aerial photos that upslope migration of the fan apex also occurred between 1953 and 1979.

8.12   Previous Studies

Tiger Wash has been site of scientific investigation since the early 1960's, when the USGS first established its crest-stage gauge. Hjalmarson and Kemna [1991] included Tiger Wash in a list of distributary flow areas in southwestern Arizona, and assigned it the highest flood hazard category, based on measurement of map-based physiographic characteristics. CH2M HILL [1992] identified Tiger Wash as one of four potential alluvial fan gauging sites for the FCDMC. The CH2M HILL study included detailed field reconnaissance, preliminary mapping of the surficial geology, descriptions of channel characteristics, mapping of historical channel changes (or lack thereof), mapping of flood high water marks, paleoflood analysis of pre-gauge flood maximums, hydrologic modeling of the watershed, detailed soil profile descriptions of the active portion of the fan, and design of a data collection system to monitor future alluvial fan flooding. Field [1994], who participated in the CH2M HILL study, describes evidence of pre-historic channel avulsions on Tiger Wash identified through field reconnaissance, soil trenching, and aerial photograph interpretation.

Following the 1997 flood, the flood of record on Tiger Wash, JE Fuller Hydrology and Geomorphology, Inc. (JEF) and the Arizona Geological Survey (AZGS) collaborated on several post-flood studies [JEF, 2000a; 2000b]. These studies focused on detailed mapping of the 1997 flood inundation limits, detailed mapping of the surficial geology of the alluvial fan surface, reconstruction of the 1997 flood peak at various points on the fan surface, and evaluation of the FCDMC alluvial fan flood hazard mapping methodology [FCDMC, 2003]. Elements of this collaborative work have been described in a variety of venues [c.f., Klawon and Pearthree, 2000; Pearthree et al., 2004]. The JEF study also included a floodplain delineation study using Federal Emergency Management Agency (FEMA) [2002a] and FCDMC [2003] three-stage methodology [JEF, 2000b] (Figure 8.30).

More recently, Pelletier *et al.* [2005] used and compared field based post-1997-flood inundation mapping, detailed surficial geologic mapping, rudimentary two-dimensional modeling, and satellite-image change detection methods of assessing the flood hazard on the Tiger Wash alluvial fan. Pelletier's work highlighted potential deficiencies in any single approach to predicting alluvial flood hazards, but concluded that composite methods that integrate geology, field observations, and numerical modeling produce the best results.

8.13   Discussion

The breadth of research on the Tiger Wash alluvial fan, in conjunction with the occurrence of a well-documented extreme flood, provides an opportunity to ponder larger questions relating to alluvial fan flood hazard assessment.

8.13.1   *What is an alluvial fan?*

The National Research Council [NRC, 1996] defines an alluvial fan as "a sedimentary deposit located at a topographic break, such as the base of a mountain front, escarpment, or valley side, that is composed of fluvial and/or debris flow sediments and which has the shape of a fan either fully or partially extended." That is, an alluvial fan is a landform that has specific compositional (alluvium), locational (at topographic break), and morphologic (fan shape) characteristics. A classic example of an alluvial fan landform is shown in Figure 8.31.

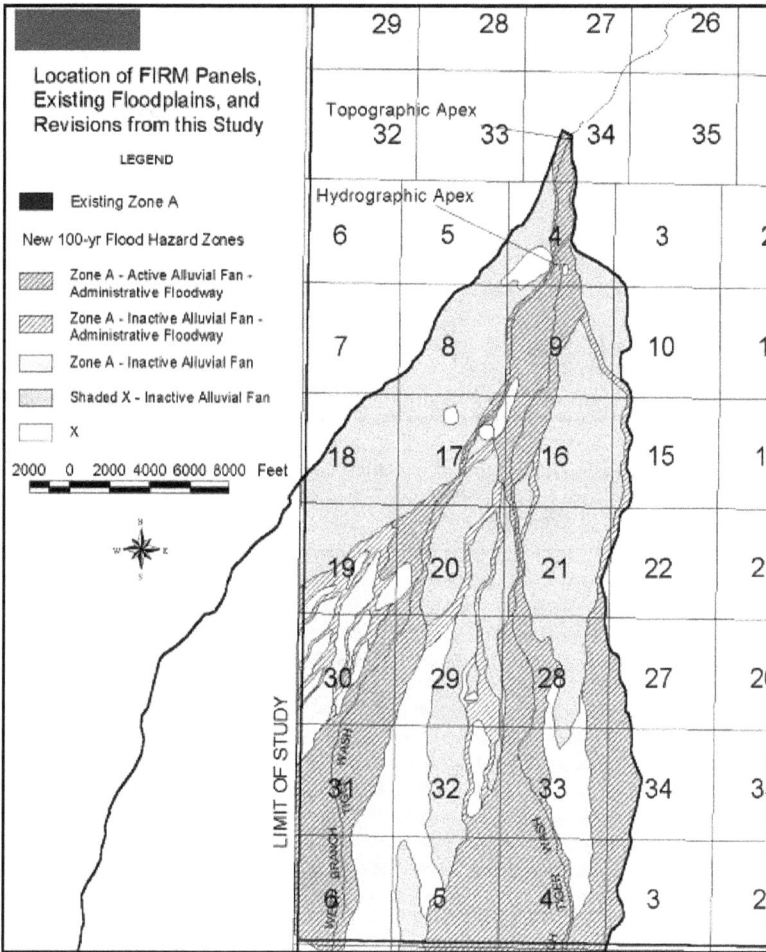

Figure 8.30   Floodplain delineation based on FEMA three-stage methodology.

The Tiger Wash site is composed of fluvially-deposited sediment, and thus meets the first of the NRC criteria without question. The landform is located near a topographic break between mountain slopes and the piedmont plain, although it lies below a valley between two mountain ranges rather than directly at a mountain front. The NRC report notes that a topographic break may be a lateral feature whereby flow is no longer contained by the main channel, which certainly occurs as Tiger Wash flows out onto the piedmont. Thus, Tiger Wash meets the second element of the

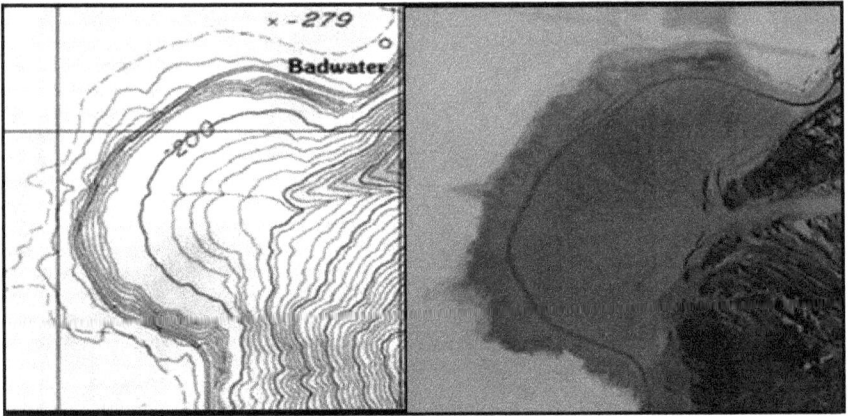

Figure 8.31   Classic alluvial fan near Badwater, Death Valley, California, USA.

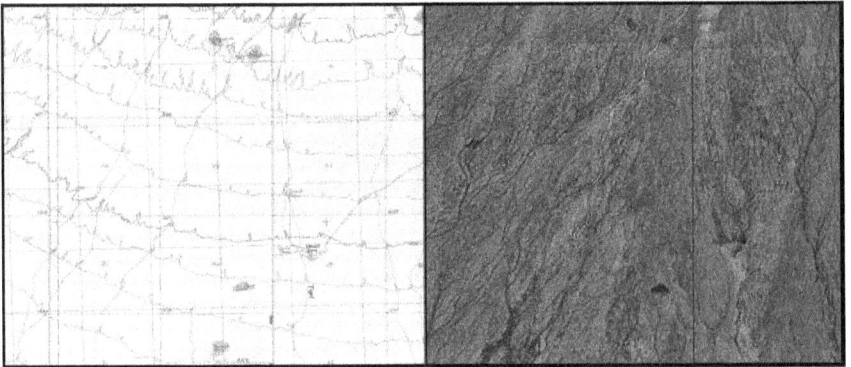

Figure 8.32   Radial contours and drainage pattern at Tiger Wash.

NRC definition. Tiger Wash has a weakly-defined, partially-extended fan shape, rather than the classic fan shape (Figure 8.32). A slight bulging of contours is visible on the topographic map, although the "fan" shape is better expressed by the radiating, distributary drainage pattern than the topography.

The differences between the Tiger Wash alluvial fan and classic alluvial fan features highlight the fact that natural landforms exist within a wide spectrum of characteristics that reflect variance in geology, physiographic setting, climate, and evolutionary stage. In some cases, landforms may have characteristics of more than one landform type. The muted fan shape at Tiger Wash as indicated by topographic contours is due primarily to the moderate slope of the fan itself, as well as to the sediment supply implications of the slight difference in fan slope and the watershed slope above the apex. In other places, the classic fan shape can be obscured by the presence of coalescing fans. Certainly, the Tiger Wash site is not a riverine, mountain, delta, or playa landform, leaving few reasonable choices if it is not an alluvial fan. With respect to flood hazard assessment, the landform type may be more of an academic, rather than practical, question. The primary flood hazard concern is whether non-riverine, alluvial fan flood processes occur on the landform.

8.13.2   *What are the key elements of alluvial fan flooding?*

The defining elements of alluvial fan flooding, as proposed by the NRC [1996] and implemented by FEMA [2002a] include the following: (1) a flood hazard that occurs only on alluvial fan landforms; (2) flow path uncertainty below a hydrographic apex, that cannot be set aside in realistic flood risk assessments or reliable hazard mitigation efforts; (3) abrupt deposition and ensuing erosion of sediment as a stream or debris flow loses its competence to carry material eroded from a steeper, upstream source area; and (4) an environment where the combination of sediment availability, slope, and topography creates an ultra-hazardous condition for which elevation on fill will not reliably mitigate the risk. The NRC definition is used in the FEMA guidelines [2002a], although the official National Flood Insurance Program (NFIP) regulations [FEMA, 2002b] define alluvial fan flooding more generically.

The first aspect of the NRC definition highlights the importance of accurate landform delineation, although the previous discussion described issues with categorizing transitional or non-classic landforms. The second criterion, flow path uncertainty, may be the single most important characteristic of alluvial fan flooding. However, the NRC definition sets a somewhat subjective standard as to the level of uncertainty. The presence of known avulsions and distributary flow paths unquestionably demonstrates the presence of "flow path uncertainty" on the Tiger Wash alluvial fan. Whether ignoring that uncertainty results in an unacceptable flood risk assessment begs the question of what is an acceptable level of risk. Uncertainty in bed elevations and channel dimensions is routinely ignored in alluvial riverine floodplain delineations, even though sedimentation processes on alluvial rivers may result in up one meter (3 feet) of variation in predicted water surface elevation [Fuller, 2002]. The uncertainty in flow depths associated with avulsions and distributary flow paths on Tiger Wash are well within that range [Pelletier, 2005]. Clearly, with respect to alluvial fan flooding, flow path uncertainty must be interpreted to mean significant variability as to what part of the alluvial fan landform may be inundated during similar sized floods, changing levels of hazard related to formation of new channels in areas previously subject only to overbank flooding (avulsions), and uncertainty of the rate of flow reaching specific concentration points on the fan surface.

The third NRC characteristic of alluvial fan flooding, loss of transport capacity and ensuing deposition of sediment derived from an upstream source is less distinctive than it might appear at first reading. Any deposition on a fluvial system is derived from an upstream source, so upstream sediment supply is not a defining characteristic. In arid climates, many riverine systems lose competence in the downstream direction due to channel transmission losses, resulting in sediment deposition. Furthermore, transport competence with respect to sediment size, if not total volume, almost always decreases in the downstream direction. Finally, most riverine floodplains are aggrading. Therefore, the key element of the third characteristic is net, long-term sediment deposition. Net deposition on the fan surface is one of the causes of flow path uncertainty.

Active alluvial fans are aggrading landforms. While the rate of aggradation may in fact be proportional to the level of hazard, it is important to note that alluvial fans aggrade over geologic time, and may not experience significant aggradation within engineering time scales. While net aggradation may not be measurable in engineering time, the effect of aggradation on landform features is easily recognized and distinguished from degrading landforms by their channel pattern, degree of channel incision and soil development, and surficial geology.

The fourth NRC characteristic, that elevation on fill does not mitigate the risk, is the most problematic for several reasons. First, the NRC report is silent on what "reliably mitigate" means or what sort of fill was envisioned. Second, sufficiently elevated and stabilized fill will always mitigate the risk to a structure placed on the fill, no matter the degree of flood hazard. Design of such fill may not be practical or affordable, but one cannot imagine a flood hazard on an alluvial fan so severe that an engineering solution does not exist. Third, the concept of risk is based on some measure of time, probability of occurrence, and tolerance for damage. Therefore, what constitutes "reliable" is subjective.

The elevation on fill characteristic is best interpreted as the fact that setting development elevations based on existing topography is problematic on alluvial fans. It is not elevation on fill per se, but rather, the potential for today's flood elevation to change significantly due to avulsions, sediment deposition, and/or channel erosion that changes flood distributions and depths on the fan surface. Alternatively, it may refer to FEMA guidance and regulations [FEMA, 2002a; 2002b] to prevent adverse impacts to adjacent properties. Placement of fill on an alluvial fan floodplain could certainly change flow distributions, concentrate flow, and initiate unacceptable sediment and erosion processes in adjacent areas.

8.13.3 *Alluvial fan boundary delineation*

An alluvial fan is bounded by its apex (upstream), toe (downstream), and lateral limits. Past mapping projects on the Tiger Wash alluvial fan elucidate several challenges in alluvial fan boundary delineation. The hydrographic apex is the highest point on the alluvial fan where physical evidence of channel bifurcation or significant flow outside the defined channel exists [FEMA, 2002a]. Early mapping efforts along Tiger Wash [CH2M HILL, 1992] focused on a point where an actual channel bifurcation existed, and thus inadequately considered the implications of overbank flooding on flow path uncertainty. Consequently, the apex was identified downstream of the point where the major avulsions occurred during the 1997 flood. Later efforts [JEF, 2000a; 2000b] emphasized surficial geologic mapping and placed the hydrographic apex at the upstream limit of the Q_{y2} surface.

The lateral limits of an active alluvial fan are relatively easy to identify where it is bounded by topographically higher, older geologic surfaces. Boundary delineation is more challenging where topographic confinement is not present or where active fans coalesce. Evidence from the 1997 flood indicates that some topographically low Pleistocene-aged surfaces along the margins of the active fan were partially inundated, suggesting that these older surfaces are being incorporated into the active fan area. Given that some of the channel avulsions exposed now buried older surfaces in the middle of the active area, it can be inferred that burial of older surfaces is a normal part of the Tiger Wash fan evolution. Therefore, reliance on surface age alone is an inadequate method of defining the fan boundary.

Consideration of hydraulic confinement and fan evolutionary processes is needed to determine fan limits.

Judgment is also required where fans coalesce. Often a fosse exists between adjacent fans that can be used to define a lateral boundary. In other cases, differences in watershed geology create sedimentological variations that clearly define the limits between neighboring fans. At Tiger Wash, no fosse or sedimentological differences were observed. Therefore, the fan boundaries between overlapping fan areas were extended laterally by following drainage paths, and/or extrapolating the orientation of surficial units, and probably overlapped the zone of influence of the adjacent landforms.

The toe of the alluvial fan landform is often defined by an axial stream, playa, or other distinct landform. If the area subject to alluvial fan flooding does not occupy the entire landform, the downstream limit may be more difficult to define. On many alluvial fan landforms in central Arizona, secondary entrenchment along the axial stream has led to a zone of incised channels along the toe of the alluvial fan landform that are not subject to uncertain flow paths, *i.e.*, inactive alluvial fan surfaces. In these cases, while the flow paths may be stable, the flow rate reaching those channels is uncertain due to the upstream alluvial fan flooding areas. Because of the flow rate uncertainty, it is important to delineate all areas connected to the alluvial fan flooding zone.

In the case of Tiger Wash, the alluvial fan flooding area near the hydrographic apex gradually transitions into a planar surface characterized by widespread sheet flooding and minimal topographic relief. The FCDMC [2003] proposes to call these distal sheet flooding areas alluvial plains, but paradoxically defines alluvial plains as occurring on alluvial fan landforms. FEMA [2002a] and the NRC [1996] state that sheet flooding on alluvial fan landforms is alluvial fan flooding. These differences may be more semantic than practical, as clearly the distal areas are dominated by sheet flow, which is subject to variable flow distribution, but has a degree of hazard less than that of the upper fan areas. For Tiger Wash, the floodplain delineation was terminated at the Central Arizona Project, a man-made structure that spanned the landform and substantively altered the natural flooding processes.

8.13.4   *Predicting avulsions*

Data from Tiger Wash provides some insight into the poorly understood process of avulsions on fluvially-dominated alluvial fans. The two major avulsive channels that formed on the Tiger Wash Fan during the 1997 flood had a number of similarities. First, both avulsions intersect the pre-flood main channel at the outside of channel bends, at points where one would expect flow to leave the channel due to momentum effects. Second, the avulsive channels formed along the approximate alignments of minor on-fan tributary swales, suggesting that overbank flow concentration is a component in avulsion formation. Third, neither avulsive channel resulted in complete abandonment of the pre-flood main channel. Floods continue to exploit the pre-1997 channel network. Fourth, there is no evidence that the avulsions were initiated due to clogging of the pre-flood main channel either by sediment or debris. On the contrary, evidence along the main channel suggests that much of the flow that exceeded the channel capacity flowed parallel to the main channel and scoured a path immediately outside the belt of dense bank vegetation.

No evidence of other major avulsions besides those in 1997 is apparent in the 50+ years of historical aerial photographic record for Tiger Wash. The presence of abandoned channels defined by alignments of palo verde trees, that by their size must have been bank vegetation, suggests that other large avulsions have occurred within the trees' life span. Other investigators have reported soil stratigraphic and radiocarbon dating indicating an avulsion occurred within approximately 650 years. Field [2001] concludes, in part from evidence at Tiger Wash that the recurrence interval for avulsions on fluvially-dominated alluvial fans in central Arizona is on the order of 50–600 years.

On Tiger Wash, the distributary channel network has a decidedly avulsive character in the mid-fan portion of the West Branch. In these areas, branching channels appear to form where channels lose capacity and mid-channel vegetation appears to clog the channel. However, it is difficult to determine whether sediment deposition, lowered channel capacity, and mid-channel vegetation are the cause or the effect of the branching channel network.

Evidence from Tiger Wash indicates that major avulsions are rare occurrences, with a return period that may exceed 100 years. Avulsions on fluvially-dominated fans may tend to occur in somewhat predictable patterns, in areas where overbank flow occurs and is concentrated along pre-existing drainage alignments. When avulsions occur, it is unlikely that the pre-flood channel network is immediately abandoned, and that the shift of flood activity to new parts of the fan occurs over long time spans.

8.13.5 *Importance of infiltration and attenuation*

Basic principles of hydrologic modeling and common sense suggest that infiltration and attenuation would play major roles in the hydrology of alluvial fans. Floods on alluvial fans originate in confined canyons with no significant overbank storage area and shallow bedrock that prevents infiltration. When floods leave the confined canyons and reach the alluvial fan, they inundate broad areas at relatively shallow depths. The broad unconfined fan surfaces not only store large quantities of flood water, they provide an extensive surface area over which infiltration can occur. Most active alluvial fan surfaces in central Arizona are composed of relatively permeable sandy alluvium, which in the arid west, are typically in a moisture deficit condition, and are capable of absorbing floodwater at relatively high rates. Floods in the arid west tend to be flashy, low volume to peak events that are susceptible to attenuation as floodwater spreads over the fan surface. Groundwater recharge is thought to occur at mountain fronts. Therefore, it is not surprising that several lines of field evidence from Tiger Wash highlight the importance of infiltration and attenuation in the hydrology of alluvial fans.

First, the channel morphology suggests decreased flow rates along any single flow path. Near the toe of the fan, there are few defined channels and very few channels that are linearly connected to the fan apex. Because channel size is directly related to flow frequency, the decreased channel size indicates decreased peak flow rates across the fan surface, *i.e.*, attenuation. Second, CH2M HILL [1992] described evidence of a large flood that completely infiltrated on the fan surface over a linear distance of about 6.4 km (4 mi), well before reaching the toe of the fan. The flood peak was estimated at approximately 56.6 cms (2,000 cfs) at Eagle Eye Road upstream of the West Branch apex. Large flotsam mats that spanned the flooded distributary channels indicated the terminus of the flood on the fan surface. These observations suggest an order-of-magnitude average infiltration rate of 33 mm/hr (1.3 in/hr), only slightly higher than the saturated hydraulic conductivity estimates for the loamy sand/sandy loam soil types on the fan surface. Third, JEF [2000a; 2000b] used slope-area methods to estimate 1997 peak discharges along the main channel branches of the Tiger Wash alluvial. Not only do the estimated peaks decrease in the downstream direction, but the sum of peaks from all branches across the fan decrease.

Some caution should be applied when including infiltration in hydrologic modeling on alluvial fans. The affect of antecedent rainfall on peak discharge is well documented. Therefore, for flood hazard assessments, it may be prudent to assume saturated conditions when estimating the infiltration rate. For fans with large surface areas relative to the watershed area, rainfall on the fan surface itself may be an important component of the flood hydrograph. Likewise, it is important to account for tributaries that flow onto the fan below the primary apex. Not all of the flood may be generated in the above-apex watershed.

8.13.6   *Flood hazard delineation*

The effective floodplain delineation of Tiger Wash [JEF, 2000a; 2000b] was completed using FEMA's three-stage methodology, which is based on interpretation of the surficial geology, use of historical data, and field observations. The three-stage methodology is a vast improvement over previous alluvial fan floodplain delineation methods [Dawdy, 1979] previously mandated by FEMA [Pelletier, 2005]. A number of valuable lessons were learned through the application of the three-stage methodology to Tiger Wash and the site work that followed the 1997 flood. First, the surficial geology provides the best depiction of the extent of the historical floodplain as well as the types of flooding that occur on the surface. Second, the location of the 1997 avulsions indicate that the channel pattern is a less reliable tool than the surficial geology, since no defined channels may exist on young surfaces that are vulnerable to flooding and avulsions. Third, topographic containment is an important requirement for defining flood boundaries. Because active alluvial fans are aggrading landforms, older surfaces will become integrated into the floodplain with time. Finally, the changes wrought by the 1997 flood highlight an unstated, but essential difference between riverine and alluvial fan floodplain delineations. On alluvial fans, the delineation is not merely a depiction of the area inundated by the next 100-year flood, it is a depiction of areas at risk of flooding over an undetermined period of time by an undetermined number of floods, as well as the impact on flood risk due to surfaces changes brought about by those floods.

It is likely that the effective floodplain delineation could be improved by incorporation of engineering/numerical modeling of the alluvial fan together with the surficial data. The primary weaknesses in numerical methods are reliance on existing condition or pre-flood topographic data, ignoring the potential for channel change, avulsion and surface aggradation, and the inability to explicitly consider accurate depictions of flow path uncertainty. The primary weaknesses in methods based solely on the geology are the potential to include flood frequencies beyond the 100-year recurrence interval, as well as the lack of numerical results often needed for engineering design. As documented in other chapters of this book, improvements in both numerical models and surficial dating methods have the potential to significantly improve the state-of-the-art with respect to alluvial fan flood hazard assessment.

8.14   Summary

The flood hazards on the Tiger Wash alluvial fan have been studied for nearly two decades. The occurrence of a well-documented major flood provided opportunities to consider a range of floodplain delineation techniques. Pelletier [2005] applied a variety of floodplain delineation techniques to the Tiger Wash fan and concluded that composite methods, that combine numerical and physical approaches, produce the most reliable results. Indeed, at Tiger Wash, our understanding of alluvial fan flood processes has been enhanced by the level of cooperation and communication between engineers, geologists, and planners.

References

Capesius, J.P. and Lehman, T.W. (2002). Determination of channel change for selected streams, Maricopa County, Arizona, USGS Water Resources Investigation Report 2001–4209, p. 63.

CH2M Hill. (1992). Alluvial Fan Data Collection and Monitoring Study: Tempe, Arizona, CH2M Hill and R.H. French, Ph.D., P.E. Consulting Engineer, Report to the Flood Control District of Maricopa County, p. 204.

Dawdy, D.R. (1979). Flood frequency estimates on alluvial fans, *Journal of the Hydraulics Division*, 105(11), 1407–1413

Demsey, K.A. (1989). Geologic map of Quaternary and upper Tertiary alluvial in the Phoenix south 30′ × 60′ quadrangle, Arizona (1:100,000), Arizona Geological Survey Open File Report 89–17.

FEMA (2002a). Guidelines for Mapping Partners and Study Contractors, Appendix G: Guidelines and Specifications for Alluvial Fans.

FEMA (2002b). Code of Federal Regulations, Title 44, Chapter 1, Part 59.A Definitions.

Field, J.J. (2001). Channel avulsion on alluvial fans in southern Arizona: Geomorphology, 37, 93–104.

Field, J.J. (1994). Surficial processes, channel change, and geological methods of flood-hazard assessment on fluvially dominated alluvial fans in Arizona, Ph.D. Dissertation, University of Arizona Department of Geosciences, p. 258.

Flood Control District of Maricopa County (2003). Piedmont flood hazard assessment for floodplain management for Maricopa County, Arizona, user's manual (draft).

Fuller, J.E. (2002). Floodplain and floodway delineation using non-traditional techniques, Proceedings of the Annual Conference of the Association of State Floodplain Managers, Albuquerque, New Mexico.

Hjalmarson, H.J. and Kemna, S.P. (1991). Flood hazards of distributary-flow areas in southwestern Arizona: USGS Water-Resources Investigation Report 91–4171, p. 58.

JE Fuller Hydrology and Geomorphology, Inc. (2000a). Inundation mapping of the September 1997 flood, surficial geologic mapping, and evaluation of the Piedmont Flood Hazard Assessment Manual for portions of the Tiger Wash piedmont: Report to the FCDMC, Contract 98-48.

JE Fuller Hydrology and Geomorphology, Inc. (2000b). Approximate floodplain delineation study for portions of Tiger Wash piedmont, technical documentation notebook: Report to FCDMC, Contract 98-48.

Klawon, J.E. and Pearthree, P.A. (2000). Field guide to a dynamic distributary drainage system: Tiger Wash, western Arizona: Arizona Geological Survey Open File Report 00-1, p. 34.

National Research Council (1996). Alluvial Fan Flooding: Washington, D.C., National Academy Press, p. 172.

Pearthree, P.A., Klawon, J.E. and Lehman, T.W. (2004). Geomorphology and hydrology of an alluvial fan flood on Tiger Wash, Maricopa and La Paz Counties, west-central Arizona: Arizona Geological Survey Open-File Report 04–02, p. 40.

Pelletier, J.D., Mayer, L., Pearthree, P.A., House, P.K., Demsey, K.A., Klawon, J.E. and Vincent, K.R. (2005). An integrated approach to flood hazard assessment on alluvial fans using numerical modeling, field mapping, and remote sensing, GSA Bulletin, 117(9/10), 1167–1180.

Pope, G.L., Rigas, P.D. and Smith, C.F. (1998). Statistical Summaries of Streamflow Data and Characteristics for Selected Streamflow-Gaging Stations in Arizona Through Water Year 1996: U.S. Geological Survey, Water Resources Investigations Report 98–4225: Tucson, Arizona, p. 907.

Shafiquallah, M., Damon, P.E., Lynch, D.J., Reynolds, S.J., Rehrig, W.A. and Raymond, R.H. (1980). K-Ar geochronology and geologic history of southwestern Arizona and adjacent areas, in Jenney, J.P. and Stone, C., (eds.), Studies in western Arizona, Arizona Geological Society Digest, 12, 201–260.

Chapter 9

Future Directions

Flood hazard identification and mitigation on alluvial fans has, from a qualitative viewpoint, a long history. Quantitatively it has a rather short history, beginning approximately in the late 1970's when the U.S. Federal Emergency Management Agency began to address flood hazard on these landforms. As this book illustrates, a need exists for both applied and fundamental research to advance this area of hydraulic/hydrologic engineering. In the following sections, some topics for future endeavor are raised and discussed.

9.1   Introduction

The preceding chapters have briefly covered a substantial body of knowledge. It is pertinent to observe that this chapter is not an original chapter by any single author, but rather a collaborative chapter written, reviewed, commented on, and added to by all the distinguished authors of this volume. Typically, a chapter such as this one would focus on an academic research agenda; however, the authors have chosen not to take this approach because it would likely not be productive. That is, the problems and issues discussed in this volume are too broad and complex to be addressed by any single governmental, academic or private sector entity. Rather, what is needed is broad co-operation and consensus between all of these diverse components of hydraulic/hydrologic engineering research. This volume notes both the contributions of the private and public sectors and of academia to addressing the alluvial fan flood hazard problem, as this is a proper and appropriate mixture of expertise, which will likely and should continue in the future. As engineers and scientists interested in this area of study, we need to start by asking "what do we know" and "what we do not know," and end by identifying cost-effective methods to gain the knowledge we believe is still needed.

9.2   What We Know — What We Don't Know

9.2.1   *Education*

United States undergraduates in civil engineering are generally required to take a course in geology; and in many cases, that course represents their last academic contact with the geoscience community. The content of many entry-level geology courses offered to undergraduate students has remained essentially static for the past three decades, covering traditional geologic concepts such as rock identification and stratigraphy, touching on environmental geology, and skipping geotechnical aspects of geology, paleohydrology, and aqueous geochemistry. Some undergraduate engineering programs have worked with the geoscience programs to develop specific engineering-related geologic courses focused on these latter engineering principles. As civil engineers, regardless of their sub-discipline of practice, their professional careers will be intertwined with colleagues in the geoscience community. It is pertinent to observe that neither undergraduate nor graduate geoscience students are typically required (or encouraged) to take courses in civil engineering, even including those who conduct water-based investigations such as hydrology or fluvial geomorphology. Therefore, improvements to the interdisciplinary relationship at the undergraduate education level are needed. The situation is exacerbated at the graduate level because of the academic limitations on taking courses outside the department or college in which the student is registered. For example, there are a wide variety of graduate-level geoscience courses that would be valuable for engineers, and there are many excellent geoscience books that could be introduced that would prove valuable to engineers [*e.g.*, Cooke and Warren, 1973 and House *et al.*, 2002]. The reverse is also true, with specific sub-disciplinary geoscience graduate students benefiting from civil engineering courses and texts.

Much of the schism between the geosciences and engineering communities is "culturally" based, being derived in part from the subordinate position many geologists of past decades felt they were placed by the engineering community. A strong focus on rigorous quantitative training of geoscientists in recent decades has resulted in improved communication and collaboration between these closely related fields, and has proven a critical step in the advancement of the study of alluvial fan flood hazards.

9.2.2 *Precipitation and flow data issues*

In semi- and arid regions rain gages are sparse and usually at low elevations, and flow gages are even rarer. Further, the periods of record for both types of gages are usually both short and sporadic. It is observed that both precipitation and flow data are more prevalent in urban and near-urban areas, but many flood mitigation and other facilities are designed and built in areas that are located along the urban fringe or far removed from existing urban areas. Further, watersheds tributary to these sites often extend into much higher elevations where there are very few gages. Given the foregoing observations, several remedies can be suggested:

(1) Whereas precipitation data are scarce, such data are more plentiful than flow data; and there are sufficient data to be interpolated and provide useful input for rainfall-runoff models [*e.g.*, Miller *et al.*, 1973; NOAA Atlas 14 HDSC Precipitation Data Frequency Server (www.hdsc.news.noaa.gov)]. Of course, the accuracy of the interpolated data relies heavily on the accuracy and density of the input data. As noted, precipitation gages and data are sparse at high elevations, and the issues resulting may extend down gradient. Resolving these issues is difficult and expensive (more gages, maintenance, monitoring, and analysis).

(2) Radar (*e.g.*, NEXRAD) is a useful tool for expanding the precipitation data coverage of traditional gages, though challenges exist to the practical use of this data. For example, it is often difficult to use radar data to estimate peak precipitation due to signal attenuation in intense rainfall events. Also, radar shadows in mountainous terrain lead to data gaps. To be truly useful in rainfall-runoff modeling, the radar data require calibration. Certainly, this is a possibility if the radar data are archived, calibrated, and made easily available. The use of radar to measure precipitation levels over large undeveloped areas is certainly an area worthy of continued research and effort.

(3) Whereas more flow gages would be useful, they are typically even more expensive overall than precipitation gages, and record even less data, as not every recordable precipitation event results in a flow event. Furthermore, large events tend to damage, destroy, overwhelm or bypass stream gages, often rendering them of little value in the measure of major flood events.

(4) Remotely sensed aircraft and satellite data continues to offer promise for many hydrologic variables, such as measuring soil moisture content, standing water volumes, land use types, vegetation cover, and both long-term and acute changes observed in watersheds such as urbanization or wildfires.

(5) Whereas the U.S. federal government maintains databases of "official" precipitation and flow data, often these databases are not shared between agencies, even within the same hydrologic basins. In addition, there are large quantities of "unofficial" and expurgated data collected within these same basins that are not included in any of these publicly accessible databases. Regionally-centralizing these databases by assigning responsibility to one agency would greatly expand the data available for analysis.

Thus, as is the case whenever engineering must interface with the natural environment, there are (sometimes substantial) data lacks and gaps; and there is a justifiable call for the deployment of additional monitoring equipment. It is suggested rather than requesting the deployment of new monitoring systems, attention should be focused on judiciously and cost-effectively supplementing existing systems, placing new monitoring equipment in critical locations, and centralizing databases from multiple entities. The issue of data is also treated in the following section.

9.2.3 *Geology and geomorphology*

The geological and geomorphological characterization of alluvial landforms, a key component of flood hazard evaluations, often involves an assessment of small elevation differences between adjacent alluvial surfaces, as well as of (often subtle) eolian, tectonic, and human processes. Such detailed geomorphic data is provided by high-resolution LiDAR imagery [Shaller *et al.*, 2004; Le *et al.*, 2008a]. Ever-expanding LiDAR coverage (both areal and temporal) will encourage significant improvements in geomorphic assessments of alluvial landforms in arid regions. Additionally, LiDAR datasets will soon be available to quantitatively compare alluvial landforms before and after major flood events, offering an opportunity to test current theories of avulsion, channel migration and sedimentation patterns.

Recent improvements in surface age dating techniques [Le *et al.*, 2008a, 2008b; Peryam *et al.*, 2008; Fletcher *et al.*, 2008; Lippencott *et al.*, 2008] also promise to revolutionize the geological characterization of alluvial landforms in arid and semi-arid regions. As highlighted by Shaller *et al.* [2006], accurate age dating of fan surfaces will permit a better understanding of modern flood processes, as well as the effects of tectonic activity and past climates on the development of alluvial landforms. Accurate age dating of alluvial surfaces should also greatly improve current estimates of sediment production from remote watersheds where such data is otherwise scarce or absent [Le *et al.*, 2008b].

9.2.4 *Monitoring and modeling*

In the previous section, fundamental data lacks and gaps were discussed; in this section, system data lacks are discussed; that is, rainfall and runoff are only a few critical pieces of the complete system. Scientists and engineers have often discussed fully instrumenting both undeveloped and developed alluvial fans to measure a full suite of hydrologic and hydraulic variables and parameters before, during, and after rainfall-runoff events to develop a comprehensive understanding of a complex alluvial fan flow system. Most of these discussions end when the magnitude of the undertaking is realistically considered. This is an appropriate point to note the outstanding work of the U.S. Department of Agriculture, Agricultural Research Service's Southwest Watershed Research Center, including the Walnut Gulch Experimental Watershed, near Tucson, Arizona. For more than fifty years, this entity has collected and interpreted fundamental semi- and arid regional hydrologic data.

Also, the efforts of one agency to collect engineering data specifically on alluvial fans is summarized in FCDMC [1992]; and some of the findings and mitigation plans put forward in that document have been implemented. This is the only comprehensive project that the authors are familiar with; however, there are likely others that are unknown and thus unacknowledged. This situation suggests two items worthy of note. First, the concept of instrumenting alluvial fans in various stages of development is sound and over time would produce a wealth of valuable information and data. However, in addition to diversity in stage of development on the fan, there also must be diversity in geographic location. This concept is complicated by the continuing changes in the stages of development on an alluvial fan, as the population growth in the Southwest United States creates farther expansion onto the surrounding alluvial fans. Second, there would need to be a clear understanding on what is to be accomplished. That is, is the intention to collect a long time series of data or is the intention to only collect data on one or more significant events and then decommission the system? Data collection time series are further complicated by the extended periods of drought now occurring throughout the Southwest United States. In any case, obtaining and retaining funding for such a project would be challenging. Supplementing collected data with modeling techniques may be an appropriate solution to evaluating the flood hazards.

The area of modeling must be divided into physical and numerical models. Although there have been many physical models built to study particular issues on alluvial fans, there have been few large scale models such as that described in Cazanacli *et al.* [2002], which were capable of modeling alluvial fans on all time scales. The laboratory described in this paper is an unique facility for conducting both basic and applied hydraulic research. Research involving alluvial fans and playa lakes is usually conducted in the field and in response to a particular problem. Unfortunately, much of this work is hidden in engineering design reports that are not widely disseminated beyond the agency or client for which the work was performed. The issue of publication is a common problem, which defies solution. Academics generally only publish in learned, peer-reviewed journals, which have very small audiences whereas professionals seldom publish beyond internal project reports. This situation is detrimental to the advancement of the profession. The area of physical modeling should be pursued as resources allow. It is worth noting that often development creates unique opportunities for experimentation. For example, Schick *et al.* [1997] describe the data and results of a flood simulation performed on an alluvial fan in the urban fringe area of Eilat, Israel — a unique field experiment that produced excellent data.

This book has extensively discussed numerical modeling, which compared with the available data is relatively advanced. At present, models are used to both forecast new events and analyze past events. As discussed in this volume the FLO-2D model, properly programmed, can do an excellent job of analyzing past events, such as the La Conchita landslide. The models that are the focus of previous chapters are single event models as are needed for flood hazard identification and mitigation; however, the earliest numerical model found was a continuous, process-based model including tectonic uplift, sedimentation, fluvial and non-fluvial flows, and random paths of motion [Hooke, 1965 and 1967]. It is perhaps time to consider constructing a new continuous simulation model and calibrate it with both field and laboratory data; that is, build a virtual alluvial fan. Such a model should have the capability to incorporate stochastic in-flood processes (such as avulsion and the incision of new channels) that allow the floodplain to be modified by surface flows in the course of a single flood event. The lack of this capability is a major limitation of current numerical models of surficial flow on alluvial floodplains. In addition, most models are constrained by data availability, specifically parameters to characterize watershed properties that influence the flow hydraulics.

9.3   Conclusion

Progress has been made in understanding hydraulic processes on alluvial fans; however, we still have a very incomplete understanding of such a complicated geologic and hydraulic landform. We have attempted to outline in this chapter a number of programs for consideration to aid in the understanding of alluvial fans. Development and re-development on alluvial fans in semi- and arid regions will continue; therefore, there will be a continuing need for flood hazard identification and mitigation as development progresses. Thus, the need for further work in this area is driven not only by the desire of engineers and scientists to understand the physical processes that occur on alluvial fans, but by the need of society to cost-effectively protect life and property as populations continue to expand into semi- and arid regions.

References

Cazanacli, D., Paola, C. and Parker, G. (2002). "Experimental steep braided flow; Application to flooding risk on fans." *Journal of Hydraulic Engineering*, 128(3), 322–330.

Cooke, R.U. and Warren, A. (1973). "Geomorphology in deserts." University of California Press, Berkeley, CA.

FCDMC, (1992). Alluvial fan data collection and monitoring study. Prepared by: CH2M-Hill, Tempe, A.Z. and R.H. French, Las Vegas, N.V. For: Flood Control District of Maricopa County, Phoenix, AZ.

Fletcher, K.E., Rockwell, T.K., Sharp, W.D. and Le, K. (2008). "Slip rates of the Elsinore fault in the southern Coyote Mountains determined by 230Th/U dating of pedogenic carbonate in offset landforms." in Cross-correlation of Quaternary dating techniques, slip rates, and tectonic models in the western Salton Trough, Friends of the Pleistocene Annual Fieldtrip, Guidebook and Roadlog, p. 12.

Hooke, R. LeB. (1965). "Alluvial fans." Ph.D. Dissertation, California Institute of Technology, Pasadena, CA.

Hooke, R. LeB. (1967). "Processes on arid-region alluvial fans." *American Journal of Science*, 266, 609–629.

House, P.K., Webb, R.H., Baker, V.R. and Levish, D.R. (2002). "Ancient flood, modern hazards: Principles and applications of paleoflood hydrology." American Geophysical Union, Washington, D.C.

Le, K., Rockwell, T., Owen, L.A., Oskin, M., Lippincott, C., Caffee, M.W. and Dortch, J. (2008a). "Late Quaternary slip rate gradient constrained from LiDAR imagery and 10Be dating of offset landforms on the southern San Jacinto Fault, California," in Cross-correlation of Quaternary dating techniques, slip rates, and tectonic models in the western Salton Trough, Friends of the Pleistocene Annual Fieldtrip, Guidebook and Roadlog, p. 13.

Le, K., Dorsey, B., Oskin, M., Housen, B. and Peryam, T. (2008b). "Climate of tectonic control on progradation of the Hueso Fm: Integration of stratigraphy, paleomagnetism and 10Be paleo-erosion rates." in Cross-correlation of Quaternary dating techniques, slip rates, and tectonic models in the western Salton Trough, Friends of the Pleistocene Annual Fieldtrip, Guidebook and Roadlog, p. 1.

Lippincott, C.K., Rockwell, T.K. and Owen, L.A. (2008). "Optically stimulated luminescence dating littoral sediments in the Imperial Valley, southern California: the methods, applicability, and problems." in Cross-correlation of Quaternary dating techniques, slip rates, and tectonic models in the western Salton Trough, Friends of the Pleistocene Annual Fieldtrip, Guidebook and Roadlog, p. 6.

Miller, J.F., Frederick, R.H. and Tracey, R.J. (1973). "Precipitation-frequency atlas of the western United States." National Oceanic and Atmospheric Administration, Washington, D.C.

National Oceanic and Atmospheric Administration (NOAA) (2009). Hydrometeorological Design Studies Center, http://hdsc.news.noaa.gov

Peryam, T.C., Dorsey, R.J., Bindeman, I., Housen, B. and Palandri, J. (2008). "Preliminary Pliocene-Pleistocene Stable-Isotope and Paleosol Data From the Fish Creek- Vallecito Basin, Southern California: Insights Into Paleoclimate From Pedogenic Carbonate." American Geophysical Union, Fall Meeting 2008, abstract #T11A-1842.

Schick, A.P., Grodek, T. and Lekach, (1997). "Sediment management and flood protection of desert towns: The effects of small catchments." Proceedings of Rabat symposium 56, IAHS PB. 245.

Shaller, P.J., Hamilton, D., Doroudian, M., Shrestha, P.L., Lyle, J.E. and Cattarossi, A. (2004). "Interpretation of tectonic, fluvial and eolian landforms in the upper Coachella Valley, California, using aerial photography, DEM and LiDAR technology." Geological Society of America, Abstracts with Programs, 36(5), p. 299.

Shaller, P.J., Hamilton, D., Shrestha, P.L., Lyle, J.E. and Doroudian, M. (2006). "Investigation of flood and debris flow recurrence — Andreas Canyon, San Jacinto Range, Southern California." ASCE World Environmental and Water Resources Congress, 2006, p. 7.